Geological Environment Monitoring and Early Warning Systems

Geological Environment Monitoring and Early Warning Systems

Guest Editor

Aidi Huo

MDPI

Basel • Beijing • Wuhan • Barcelona • Belgrade • Novi Sad • Cluj • Manchester

Guest Editor
Aidi Huo
School of Water and
Environment
Chang'an University
Xi'an
China

Editorial Office
MDPI AG
Grosspeteranlage 5
4052 Basel, Switzerland

This is a reprint of the Special Issue, published open access by the journal *Sustainability* (ISSN 2071-1050), freely accessible at: https://www.mdpi.com/journal/sustainability/special_issues/ZSTPZWR0L0.

For citation purposes, cite each article independently as indicated on the article page online and as indicated below:

Lastname, A.A.; Lastname, B.B. Article Title. *Journal Name* **Year**, *Volume Number*, Page Range.

ISBN 978-3-7258-5573-5 (Hbk)
ISBN 978-3-7258-5574-2 (PDF)
https://doi.org/10.3390/books978-3-7258-5574-2

Cover image courtesy of Aidi Huo

Contents

About the Editor

Aidi Huo

Aidi Huo is a professor at the School of Water and Environment at Chang'an University, Xi'an City, Shaanxi Province, China. He is mainly engaged in research on soil erosion in arid areas, hydrological ecology, geological disasters, watershed hydrological models, and extreme hydrological events under climate change. In recent years, he has collected extensive data in the fields of geological disasters and ecological environment governance on the Loess Plateau, as well as in the research of loess hydrogeology. He has also made a series of significant research achievements in the areas of simulation of the water cycle in river basins, evolution of the ecological environment, and physical model experiments of loess. He has presided over 30 various scientific research projects at the national and provincial levels, including major national natural science foundation projects and face projects of the National Natural Science Foundation, as well as key research projects of Shaanxi Province. He has published 3 academic books and over 100 academic papers as the first author or corresponding author, among which more than 40 are indexed by SCI. He is the first inventor of 15 national invention and utility model patents, and also holds 1 PCT international patent and 2 software copyrights. Currently, he is a member of the International Hydrology Science Association and the Shaanxi Ecological Society.

Preface

Today, the planet we live on is facing numerous complex and interconnected challenges, and many of these problems in the geological environment are of great concern, hence the need for this Reprint, which focuses on "Geological Environment Monitoring and Early Warning Systems".

As an important means to understand the "health status" of the earth, geological environment monitoring delivers an accurate physical examination, which can capture the subtleties of environmental changes in real time and provide a reliable basis for subsequent decision-making. The early warning mechanism is a protective net built on this basis, allowing us to prepare for possible geological disasters and other crises in advance and protect the safety of life and property.

Soil erosion is a process affected by both natural factors and human activities, which is constantly changing the face of the earth and threatening the sustainable use of land. As a powerful external force affecting the geological environment, human activities have positive and negative effects, and analyzing the relationship between them and various elements of the geological environment will help guide us to a more scientific research path of soil erosion.

Hydrological models are key to deciphering hydrological processes, helping us to understand the distribution, flow, and coupling mechanisms of water resources with the geological environment. They can accurately reproduce complex hydrological phenomena through mathematical expressions and computer simulations based on different geographical conditions, meteorological data, and other elements. Whether it is the change in river water volume or the water balance in the region, hydrological models can effectively quantitatively analyze and predict outcomes, provide strong theoretical support for the rational allocation of water resources, environmental monitoring, ecological restoration, and many other aspects, and play an indispensable and important role in the research and practice of the entire geological environment system.

The Reprint brings together the wisdom of many experts and scholars, embodying their deep thinking in field research, experimental analysis, and theoretical derivation. We sincerely hope that through the presentation of this Reprint, we can build a platform for extensive exchange of different views so that more scientific researchers and practitioners can draw inspiration from the insights provided within, all the while contributing to promoting the development of geological environment-related fields and realizing the harmonious coexistence of man and nature.

Aidi Huo
Guest Editor

sustainability

MDPI

Article

Spatiotemporal Variation in the Coupling Relationship between Human Activities and Soil Erosion—A Case Study in the Weihe River Basin

Zhixin Zhao [1,2], Aidi Huo [1,2,3,*], Qi Liu [1,2], Jianbing Peng [4], Ahmed Elbeltagi [5], Mohamed EL-Sayed Abuarab [6] and Mohamed Said Desouky Abu-Hashim [7]

1 School of Water and Environment, Chang'an University, Xi'an 710054, China; zhao1201xin@163.com (Z.Z.); liuaqqi@chd.edu.cn (Q.L.)

2 Key Laboratory of Subsurface Hydrology and Ecological Effects in Arid Region, Ministry of Education, Chang'an University, Xi'an 710054, China

3 Key Laboratory of Mine Geological Hazards Mechanism and Control, Ministry of Natural Resources, Xi'an 710054, China

4 College of Geological Engineering and Geomatics, Chang'an University, Xi'an 710054, China; dicexy_2@chd.edu.cn

5 Agricultural Engineering Department, Faculty of Agriculture, Mansoura University, Mansoura 35516, Egypt; ahmedelbeltagy81@mans.edu.eg

6 Agricultural Engineering Department, Faculty of Agriculture, Cairo University, Giza 12613, Egypt; mohamed_abuarab@cu.edu.eg

7 Soil Science Department, Faculty of Agriculture, Zagazig University, Zagazig 44519, Egypt; mabuhashim@zu.edu.eg

* Correspondence: huoaidi@chd.edu.cn

Abstract: Studying the relationship between human activities and soil erosion on a regional scale is of great significance for macro-decision-making in soil erosion prevention and control. The entropy weight method and RUSLE model are used to analyze the spatiotemporal variation in human activity intensity (HAI) and soil erosion in the Weihe River Basin (WRB) from 2005 to 2020. Through geographic detectors and a four-quadrant model, the impact of various driving factors and the coupling degree of the human–land relationship are studied. The results showed: (1) During the past 15 years, the moderate, high, strong, and severe erosion areas in the WRB decreased by 9.88%, 35.89%, 45.17%, and 70.05%, respectively. The ratio of the historical sand transport modulus to the RUSLE model result is 0.83, indicating that the results obtained by the RUSLE model can be used for further analysis. (2) Slight and weak degrees account for 80% in the northwest region. The high and strong regions are mainly distributed in the Shaanxi section, accounting for 3% of the total basin. (3) The coupling between human activities and soil erosion is constantly strengthening, and the joint effect of pop and crop is the main reason for the slowdown and spatial differences in soil erosion. This indicates that the ecological environment became stable. These findings contribute by acting as references for soil and water conservation and management in the WRB to promote a harmonious relationship between humans and the environment.

Keywords: anthropogenic activity; soil erosion; RUSLE; four-quadrant model; geographic detectors

1. Introduction

The human–land relationship is defined as the relationship between the survival and development of human society or human activities and the geographical environment [1]. In recent years, there has been a great shift in research focus to the coordinated and sustainable relationship between human beings and the living environment from land-use change and assessment [2,3]. Soil erosion is one of the major environmental problems causing worldwide soil degradation and crop yield reduction, which seriously threatens the sustainable development of society and the environment [4]. Changes in soil erosion are generally considered to be the result of a combination of natural factors and human activities. However, the effects of natural factors are generally long-lasting, while the effects of human activities are visible for a short period of time. With the enhancement of human activities and the expansion of their scope, human activities have gradually become the key factor leading to the change in soil erosion [5,6]. Research on the relationship between human activities and soil erosion changes and their spatiotemporal differentiation at a regional scale is of great significance for macroplanning of soil erosion prevention and control [7–9]. The research on the relationship between human activities and soil erosion changes and their spatiotemporal differences is a scientific issue to be solved urgently in current soil erosion research.

In the past 20 years, most of the research on soil erosion has focused on the spatial variability of soil erosion under the background of land-use or landscape pattern changes or on the use of models such as InVEST, SWAT, RUSLE, etc., to assess spatial and temporal changes in soil erosion [10–12]. For example, Zhang et al. [13] discussed the impact of land disturbance and restoration on runoff and sediment yield by comparing six temporary land protection measures. Xu et al. [14] analyzed the spatiotemporal evolution and susceptibility map of soil erosion on the Loess Plateau by using statistical methods and frequency ratio models through field investigations. Zhang et al. [3] evaluated soil erosion for different land-use types in each sub-basin of the Liusha River Basin based on the SWAT model. Li et al. [15] used the InVEST model to explore the spatial distribution characteristics of soil erosion intensity in Anxi County. However, it is not enough to only emphasize the impact of land-use change while ignoring socio-economic factors, or to only focus on the effects of different human activity factors and ignore spatial heterogeneity in the study of the human–land relationship as it affects soil erosion.

The relationships between human activities and soil erosion are closely bonded and complex, and these relationships can vary depending on the natural environment and human factors in the region [16]. It is found that human activities act on soil erosion in various ways, with complex processes and intensities, and the behavior is difficult to predict. Jia et al. [17] explored the effects of natural and socio-economic factors on soil erosion in the Weihe River Basin using a geodesic probe. Li et al. [18] used multisource data to decouple the causes of soil erosion in South China, and obtained the relative contribution of climate change and human activities. Wang et al. [19] used the coupled coordination degree model to evaluate the coupling coordination level between population demand, agricultural production, and soil erosion in 281 cities in China from 1995 to 2010, as well as the impact of socio-economic factors. However, the spatialization of human activities on soil erosion has not been reflected. The four-quadrant model can be used to indicate the interaction strength and correlation degree of factors from two dimensions [16]. It was originally used to analyze changes in the real estate market, but with the integration and development of other disciplines, it has been widely applied in multiple fields, such as culture and tourism, mining and agricultural ecology, human activities, and ecosystem services [20]. Research has shown that the four-quadrant model is suitable for the interaction between two subsystems. By quantifying the spatial distribution of human activities and soil erosion, the human–land relationship of soil erosion can be intuitively analyzed, and the areas that need to be controlled can be identified, providing targeted reference for soil and water conservation policies and work.

As the largest first-level tributary of the Yellow River Basin, the WRB undertakes an important ecosystem service function and plays a critical role in maintaining the ecological balance and social and economic development in Northwest China [21]. However, rapid urbanization and industrialization have increased the pressure on the ecological environment of the WRB, exposing it to serious problems of soil erosion and ecological degradation [22]). Several studies have revealed an improvement trend in vegetation and a good ecological transition in the WRB from 2000 to 2014 [23,24]. However, the spatiotemporal coupling between human activities and soil erosion changes should be strengthened, especially over different spaces. The human activity factors considered in such studies are relatively single, so the results are often more accurate. However, the effects of human activities on soil erosion are complex and diverse, and it is difficult to comprehensively investigate the patterns of human activities on a regional scale. In addition, research on the use of socio-economic statistical data to characterize human activities and analyze their relationship to soil erosion and spatiotemporal heterogeneity under the support of 3S technology has also made corresponding progress. Such studies often use regression models to measure the impact of soil erosion changes and human activities. However, socio-economic statistical data are limited by administrative units, and the representation of human activities is insufficient in terms of comprehensiveness and temporal and spatial details.

In this study, remote sensing and GIS technology were used as follows: (1) to integrate socio-economic data, land-use data, and topographic data; (2) to construct the human activity degree index at the raster level; (3) to simulate the soil erosion modulus in the study area using the RUSLE model; (4) to quantitatively analyze the relationship between soil erosion changes and human activity degree in the Weihe River Basin at the raster scale; and (5) to analyze the driving mechanism of human activities on soil erosion from multiple perspectives, such as population change, GDP, and land-use pattern change. On the basis of traditional research, we spatialized the relationship between human activities and soil erosion. We used a four-quadrant model to identify the spatiotemporal evolution characteristics of the coupling relationship between HAI and soil erosion, and provided targeted suggestions for controlling soil erosion in areas that need to be controlled. The main content is reflected in three aspects: (1) the quantification of the spatiotemporal variation characteristics of HAI in the WRB; (2) the clarification of the spatiotemporal changes in soil erosion in the WRB from 2005 to 2020, and the use of geographic detectors to analyze driving factors; and (3) the evaluation of the relationship between HAI and soil erosion in the WRB using a four-quadrant model, and the proposal of targeted recommendations for regional water and soil conservation. The results offer recommendations for soil and water conservation management and ecological sustainable development in the WRB.

2. Materials and Methodology

2.1. Study Area

The Weihe River belongs to the largest tributary of the Yellow River, with a total length of 818 km and a total basin area of about 1.35×10^5 km^2 (103°05′–110°05′ E, 33°50′–37°05′ N) [17]. The WRB spans 84 counties in Shaanxi Province, Gansu Province, and Ningxia Hui Autonomous Region, of which the Shaanxi, Gansu, and Ningxia sections account for 49.8%, 44.1%, and 6.1%, respectively (Figure 1). There are many geomorphic types in the basin, including the North Shaanxi, Longdong Eastern Loess Plateau, Qinling Mountains, and Weihe Valley. The region has a continental monsoon climate with an average annual precipitation of 500 to 800 mm and an average annual temperature of 7.8 to 13.5 °C [21]. The WRB is not only an essential strategic highland in the economic field in the western region of China but also a major industrial, agricultural, and energy base in China; it is also the location of an equally important transportation hub in the western region of China. However, due to its long history of indiscriminate logging, the ecology of the watershed has been damaged, resulting in the Weihe River becoming an important source of sediment for the Yellow River [23].

2.2. Data Sources

Table 1 displays a detailed description of the data sources, and the access date of the data is 1 March 2023. The data were all uniformly resampled to 1 km before use in this study to make the data additive.

2.3. Research Methodology

The RUSLE model was used to calculate soil erosion in the WRB. Four human activity influencing factors including night lighting, population density, GDP density, and the proportion of cultivated land, were selected to quantify the HAI in the WRB, and the driving factors affecting soil erosion were analyzed using a geographic detector. On this premise, a four-quadrant model of HAI and soil erosion was established to spatialize the coupling relationship between HAI and soil erosion. The research method is reported in the following subsections.

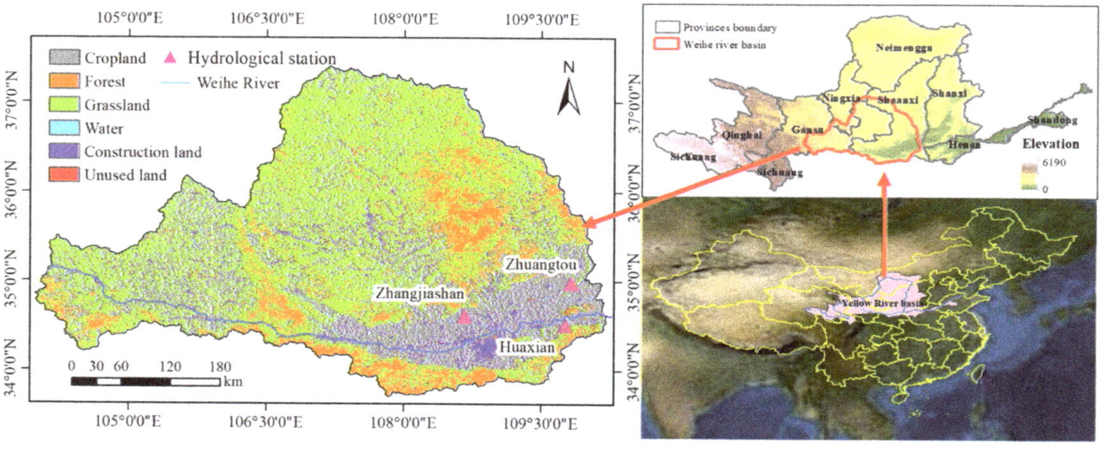

Figure 1. Geographical location, elevation, and land use of the study area.

Table 1. Detailed description of data.

Data	Resolution	Time Period	Source
DEM	90 m	2005, 2010, 2015, 2020	http://www.gscloud.cn/search
NDVI	250 m	2005, 2010, 2015, 2020	https://ladsweb.modaps.eosdis.nasa.gov
Monthly rainfall data	1 km	2005, 2010, 2015, 2020	http://www.geodata.cn/
Land use	500 m	2005, 2010, 2015, 2020	http://www.ncdc.ac.cn
Soil	1:1 million	2009	http://www.ncdc.ac.cn
Population	1 km	2005, 2010, 2015, 2020	https://www.worldpop.org/
GDP	1 km	2005, 2010, 2015	http://www.resdc.cn/
Cropland	3 km	2007, 2011, 2015, 2020	https://glad.umd.edu/dataset/croplands
PANDA	1 km	2005, 2010, 2015, 2020	http://data.tpdc.ac.cn

2.3.1. RUSLE Model

The RUSLE model is a modified version of the general soil erosion model, with a broader range of applications based on the USLE modification of the general soil erosion model [25]). The expression of the model is as follows:

$$A = R \times K \times L \times S \times C \times P \tag{1}$$

where A is the soil erosion amount in $t \cdot km^{-2} \cdot a^{-1}$; R is the rainfall erosivity factor in $MJ \cdot mm/(hm^2 \cdot h \cdot a)$; K indicates the soil erodibility factor in $t \cdot hm^2 \cdot h/(hm^2 \cdot MJ \cdot mm)$; L and S are the slope length factor and slope factor, respectively, usually expressed as LS; C is the vegetation cover and management factor; P is the soil conservation measure factor. LS, C, and P factors are all dimensionless units.

In this study, the rainfall erosion factor was calculated by annual rainfall as proposed by Zhang and Fu [26]):

$$R = \alpha_4 \times F_F^{\beta_4} \tag{2}$$

$$F_F = 1 \times N^{-1} \sum_{i=1}^{N} \left[\left(\sum_{j=1}^{12} P_{i,j}^2 \right) \times \left(\sum_{j=1}^{12} P_{i,j} \right)^{-1} \right] \quad (3)$$

where R is the multiyear average rainfall erosion force (MJ·mm·hm^{-2}·h^{-1}·a^{-1}); $P_{i,j}$ is the ith year, j is the annual month rainfall (mm), and N is the number of years; and α_4 and β_4 are model parameters with values of 0.1833 and 1.9957, respectively.

The soil erodibility factor (K) is a composite representation of the soil's resistance to erosion and was calculated with reference to previous studies [27] as:

$$K = \{0.2 + 0.3exp\left[-0.026SAN\left(1 - \frac{SIL}{100}\right)\right]\} \times \left(\frac{SIL}{CLA+SIL}\right)^{0.3} \times \left[1.0 - \frac{0.25C}{C+exp(3.72-2.95C)}\right] \times \left[1.0 - \frac{0.7SNI}{SNI+exp(-5.51+22.9SNI)}\right] \quad (4)$$

$$SNI = 1 - \frac{SAN}{100} \quad (5)$$

where K is the soil erodibility factor value in t/(MJ-mm); SAN, SIL, and CLA are the sand, powder, and clay grain-mass content (%), respectively; and C is the organic carbon mass content (%).

According to the empirical formula established by the slope of cultivated land in the United States, the formula was modified by calculating the slope length factor (L) and slope factor (S) in segments, drawing on the method of Cao et al. [28]:

$$S = \begin{cases} 10.8 \times sin\theta + 0.036 & \theta < 5^\circ \\ 16.8 \times sin\theta - 0.5 & 5^\circ < \theta < 10^\circ \\ 21.9 \times sin\theta - 0.96 & \theta > 10^\circ \end{cases} \quad (6)$$

where S is the slope factor and θ is the ground slope.

$$L = \left(\frac{\lambda}{22.13}\right)^m \quad (7)$$

$$m = \begin{cases} 0.2 & \theta < 1^\circ \\ 0.3 & 1^\circ < \theta < 3^\circ \\ 0.4 & 3^\circ < \theta < 5^\circ \\ 0.5 & \theta \geq 5^\circ \end{cases} \quad (8)$$

where θ denotes ground slope, L denotes slope length factor value, λ denotes slope length value, and m denotes slope factor index.

In this paper, the vegetation cover factor C is based on the algorithm of Almagro et al. [29]:

$$C = \sum_{i=1}^{12} exp\left[\frac{(-2) \times NDVI_{maxi}}{1 - NDVI_{maxi}}\right] \quad (9)$$

where C is the vegetation cover factor; and $NDVI_{maxi}$ is the $NDVI$ maximum value in the ith month of the study area.

Based on pertinent studies in loess hilly gully areas and combined with the actual situation of the WRB [17,30], P of 0.2, 0.6, 0.8, 0, 0, and 1 were obtained for arable land, forest land, grassland, water area, construction land, and unused land, respectively.

2.3.2. Entropy Method

Entropy has been applied in numerous research disciplines to determine the weight of index variables by assessing the degree of disorder in variables and contrasting the information content of variables [31]. Standardization is required to make each datum additive, with the positive indicator in Equation (10) and the negative indicator in Equation (11).

$$Y_{ij} = \frac{X_{ij} - X_{imin}}{X_{imax} - X_{imin}} \quad (10)$$

$$Y_{ij} = \frac{X_{imax} - X_{ij}}{X_{imax} - X_{imin}} \quad (11)$$

where X_{imax} and X_{imin} are the maximum and minimum values of the index, respectively, and Y_{ij} is the jth impact prevention and control factor of the standardized results.

$$E_j = -\frac{1}{lnm} \sum_{i=1}^{m} \left(\frac{Y_{ij}}{\sum_{i=1}^{m} Y_{ij}} \right) ln \left(\frac{Y_{ij}}{\sum_{i=1}^{m} Y_{ij}} \right) \quad (12)$$

where E_j is the information entropy of the influence factor; the smaller the value, the less variation of the factor and the greater the weight. $\frac{Y_{ij}}{\sum_{i=1}^{m} Y_{ij}}$ is the proportion of the standardized value Y_{ij} to the total standardized value.

$$W_j = \frac{1 - E_j}{k - \sum_{j=1}^{k} E_j} \quad (13)$$

where W_j is the weight occupied by the factor and k is the number of influencing factors.

2.3.3. Geographical Detector

The geographic detector is a statistical model for analyzing spatial variation, proposed by Wang et al. [31], which consists of four modules: factor detection, interaction detection, risk detection, and ecological detection. Based on the principle of spatial autocorrelation, factor detection and interactive detection can effectively determine the impact of a single independent variable and the interaction between two independent variables on the dependent variable, and can quantitatively analyze the impact of environmental change on vegetation change in the study area. The formulas are as follows:

$$q = 1 - \frac{1}{N\sigma^2} \sum_{h=1}^{L} N_h \sigma_h^2 = 1 - \frac{W}{T} \quad (14)$$

$$W = \sum_{h=1}^{L} N_h \sigma_h^2 \quad (15)$$

$$T = N\sigma^2 \quad (16)$$

where h is the hierarchy of the independent variable X, and h = 1,2, ... L; and L is the classification of the independent variable X or the dependent variable Y. The range of q values is [0, 1]. The larger the q value, the stronger the spatial heterogeneity of the dependent variable Y, and the stronger the explanatory power of the independent variable X on the dependent variable Y. N_h and N represent the number of units in layer h and the entire area, respectively, and represent the variances of layer h and region Y, respectively; and W and T represent the sum of intralayer variances and the total regional variances, respectively.

2.3.4. Four-Quadrant Model

The four-quadrant model combines dynamic and static analysis to analyze the relationship between the two factors [32]. In this study, the model was applied to the coupling relationship between human activities and soil erosion in the WRB to explore the current situation of the human–land relationship and spatial differentiation characteristics. A four-quadrant model was constructed with the intensity of human activities as the horizontal axis and the soil erosion modulus as the vertical axis. The positions of the quadrants where the index values are located reflect the degree of differences to which human–land connections differ among various assessment units (Figure 2). Based on the natural break point method in ArcGIS, the human–land relationship is divided into four zones: excellent, good, general, and poor qualities (Tables 2 and 3).

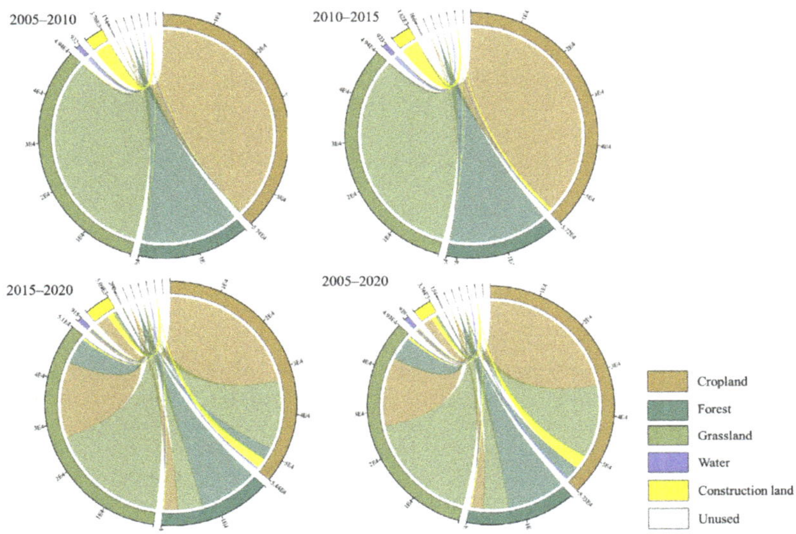

Figure 2. Chord chart of changes in different land-use types in the WRB from 2005 to 2020.

Table 2. Weight assignment results.

Indicator	Direction	Weight
Population	Positive	0.29
GDP	Positive	0.26
Cropland	Positive	0.12
PANDA	Positive	0.33

Table 3. Four-quadrant zoning table of HAI and soil erosion.

Quadrant	First Quadrant	Second Quadrant	Third Quadrant	Fourth Quadrant
HAI	0.106–0.875	0–0.106	0–0.106	0.106–0.875
Soil erosion	19,044.43–231,253.78	19,044.43–231,253.78	0–19,044.43	0–19,044.43
Human–environment interaction	Good	Poor	General	Excellent

The quadrant flow state index (*QFSI*) is established to reveal the dynamic changes in the human–land relationship using the following equation:

$$QFSI(i) = \frac{F(i)_{in} - F(i)_{out}}{F(i)_{in} + F(i)_{out}} \; (-1 \leq \text{QFST} \leq 1) \tag{17}$$

where $QFSI(i)$ is the quadrant flow state index, $F(i)_{in}$ is the *i*th quadrant shift-in area, and $F(i)_{out}$ is the *i*th quadrant shift-out area. When *QFST* is less than 0, it means the quadrant area is in the decreasing state; when *QFST* is greater than 0, it means the quadrant area is in the increasing state. When *QFST* is close to 0, it means that the quadrant is in the equilibrium state of two-way conversion.

3. Result

3.1. Changes to the Dynamics of LUCC

Figure 2 shows the changes in different land-use types in the WRB from 2005 to 2020. Cropland is the main type of land use in the WRB, accounting for 42.22% of the total area, followed by grassland (37.26%) and forest land (16.45%). The cumulative area of water and other types of land accounts for less than 5% of the total area. From 2005 to 2020, with population growth and rapid economic development, the land-use patterns in the WRB have undergone significant changes. The order of different land-use types according to the degree of area increase is grassland > construction land > unused land > water area > forest land > cropland. The area of cropland and forest land has generally decreased, with the former reaching 3.31×10^5 ha. However, the area of grassland and construction land has increased year by year, with grassland increasing the most, reaching 2.04×10^5 ha. The growth rate of construction land is the highest, with a dynamic increase of 35.39%.

3.2. Spatial–Temporal Variation Characteristics of Soil Erosion

3.2.1. Rationality and Validation of the Model

Using the annual sediment transport modulus data of hydrological stations within the watershed, spatial interpolation is performed in CIS to obtain the sediment transport modulus of the study area and compare it with the soil erosion modulus. Due to the use of multiyear average rainfall erosivity in this article, it is necessary to compare the multiyear average sediment transport modulus with the multiyear average soil erosion intensity in the watershed. The hydrological monitoring stations used are Zhuangtou, Zhangjiashan, and Huaxian (Figure 1). According to the "China River Sediment Bulletin" published on the website of the Ministry of Water Resources (http://www.mwr.gov.cn/, accessed on 1 March 2023), the average sediment transport moduli for many years in Zhuangtou, Zhangjiashan, and Huaxian are 2520, 4580, and 2680 $t{\cdot}km^{-2}{\cdot}a^{-1}$, respectively. Kriging interpolation was performed in GIS to obtain the annual average sediment transport modulus of 3523.26 $t{\cdot}km^{-2}{\cdot}a^{-1}$ in the WRB. Through calculation, the average soil erosion intensity of the Weihe River Basin from 2005 to 2020 was 4231.32 $t{\cdot}km^{-2}{\cdot}a^{-1}$, and the ratio of sediment transport modulus to the results of the RUSLE model was 0.83, indicating that the results obtained from the RUSLE model can be used for further analysis.

3.2.2. Time Variation in Soil Erosion Intensity

The soil erosion intensity of the study area in Phase 4 is divided into six erosion levels: slight, weak, moderate, high, strong, and severe, according to the Classification Standard for Soil Erosion. The area percentage of different soil erosion levels is presented in Table 4.

Table 4. Area proportions of different soil erosion levels in the WRB from 2005 to 2020.

	Slight	Weak	Moderate	High	Strong	Severe	Average Erosion Modulus
2005	51.56%	8.56%	9.38%	7.64%	9.25%	13.61%	6363.49
2010	56.13%	9.38%	8.77%	6.54%	7.92%	11.25%	5451.36
2015	68.91%	9.63%	7.33%	4.51%	4.89%	4.72%	2571.37
2020	66.72%	10.78%	8.45%	4.90%	5.07%	4.08%	2539.06

In 2005, the minimum erosion modulus was 0 $t{\cdot}km^{-2}{\cdot}a^{-1}$, the maximum erosion modulus was 214,439 $t{\cdot}km^{-2}{\cdot}a^{-1}$, and the average erosion modulus was 6363.49 $t{\cdot}km^{-2}{\cdot}a^{-1}$. Approximately 51.56% of the area had an erosion modulus of less than 1000 $t{\cdot}km^{-2}{\cdot}a^{-1}$, about 69.5% of the area erosion modulus was below 5000 $t{\cdot}km^{-2}{\cdot}a^{-1}$, and an erosion modulus of 15,000 $t{\cdot}km^{-2}{\cdot}a^{-1}$ or more accounted for about 13.61% of the watershed area.

In 2010, the minimum erosion modulus was 0 $t{\cdot}km^{-2}{\cdot}a^{-1}$, the maximum erosion modulus was 231253.78 $t{\cdot}km^{-2}{\cdot}a^{-1}$, and the average erosion modulus was 5451.36 $t{\cdot}km^{-2}{\cdot}a^{-1}$. Approximately 56.13% of the area had an erosion modulus of less than 1000 $t{\cdot}km^{-2}{\cdot}a^{-1}$, about 74.28% of the area erosion modulus was below 5000 t·km^{-2}·a^{-1}, and an erosion modulus of 15,000 t·km^{-2}·a^{-1} or more accounted for about 11.25% of the watershed area.

In 2015, the minimum erosion modulus was 0 t·km^{-2}·a^{-1}, the maximum erosion modulus was 92,797.72 t·km^{-2}·a^{-1}, and the average erosion modulus was 2571.37 t·km^{-2}·a^{-1}. Approximately 68.91% of the area had an erosion modulus of less than 1000 t·km^{-2}·a^{-1}, about 85.87% of the area erosion modulus was below 5000 t·km^{-2}·a^{-1}, and the erosion modulus of 15,000 t·km^{-2}·a^{-1} or more accounted for about 4.72% of the watershed area.

In 2020, the minimum erosion modulus was 0 t·km^{-2}·a^{-1}, the maximum erosion modulus was 231,100.56 t·km^{-2}·a^{-1}, and the average erosion modulus was 2539.06 t·km^{-2}·a^{-1}. Approximately 66.72% of the area had an erosion modulus of less than 1000 t·km^{-2}·a^{-1}, about 85.95% of the area erosion modulus was below 5000 t·km^{-2}·a^{-1}, and am erosion modulus of 15,000 t·km^{-2}·a^{-1} or more accounted for about 4.08% of the watershed area.

3.2.3. Spatial Variation in Soil Erosion Intensity

Figure 3 reveals that there was significant spatial variation in soil erosion from 2005 to 2020. The distribution of severe eroded areas was relatively stable, mostly with high and steep terrain and poor vegetation coverage, resulting in serious erosion and requiring focused targeted treatment. The areas with strong erosion were mainly distributed in the northern and western parts of the WRB, and the area gradually decreased with time changes. Areas of high erosion mostly existed around areas of strong erosion. Some areas of high erosion also existed in the central part from 2005 to 2010, and areas of high erosion showed a trend of conversion to moderate and slight erosion from 2010 to 2015.

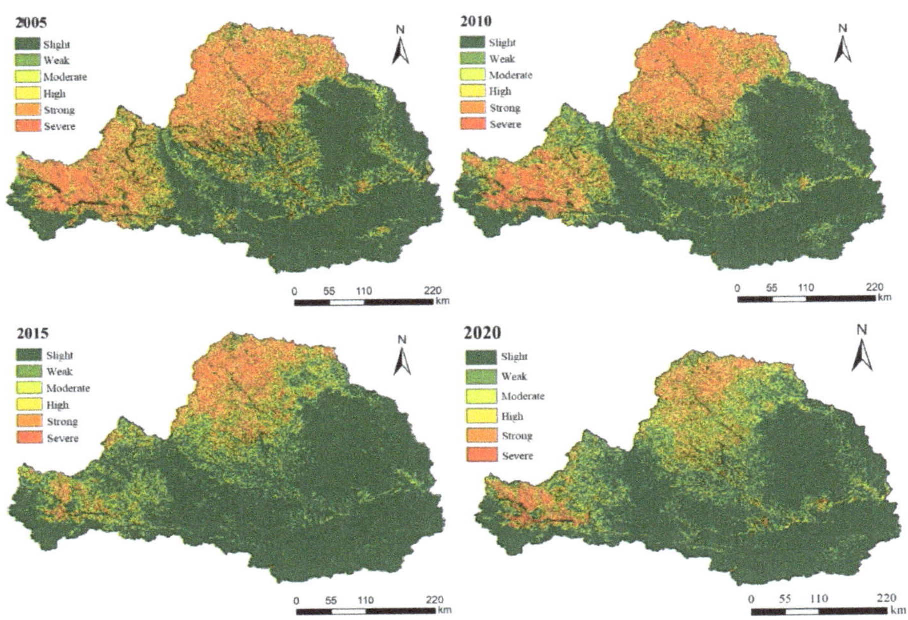

Figure 3. Spatial distribution of soil erosion from 2005 to 2020.

The change degree of the stability rate of soil erosion intensity in different years and the change intensity of soil erosion in each province are shown in Figure 4. It can be seen from Figure 4A that the stability rate of weak, moderate, high, and strong erosion in the WRB is relatively low as a whole. The stability rate of severe erosion is only higher than 50% in the period of 2005–2010, and the stability rate of severe erosion is lower than 30% during the period of 2010–2020, which indicates a significant decreasing trend of soil erosion intensity. The WRB is composed of three sections: Gansu section, Ningxia section, and Shaanxi section. The conversion rates (CVR) of weak, moderate, strong, and severe in all three sections from 2005 to 2020 were above 80%, proving that soil erosion showed a decreasing trend (Figure 4B–D).

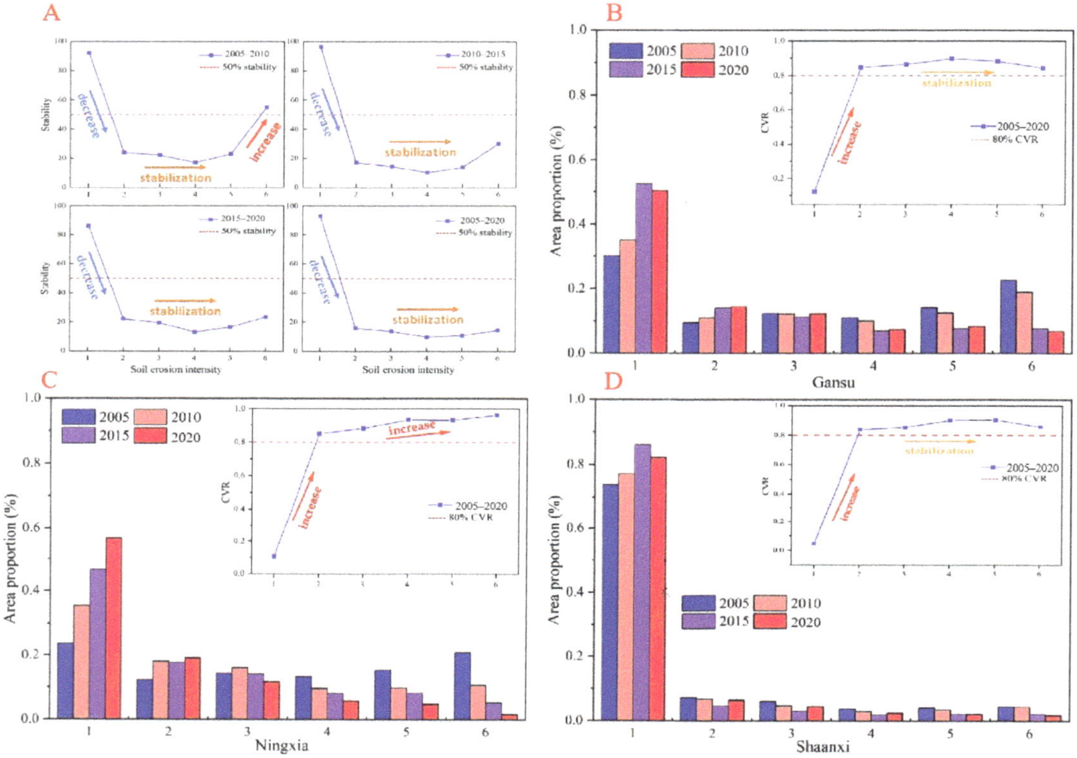

Figure 4. Degree of change in the stability rate of soil erosion intensity in different years ((**A**) represents the steady change of soil erosion in different years, (**B**) represents the change of soil erosion intensity and conversion rate in the Gansu section of the WRB, (**C**) represents the change of soil erosion intensity and conversion rate in the Ningxia section of the WRB, and (**D**) represents the soil erosion intensity and conversion rate in the Shaanxi section of the WRB rate change. Note: abscissa 1–6 represents the degree of erosion, with 1 representing slight and 6 representing severe).

3.3. Spatial–Temporal Variation Characteristics of HAI

The natural breakpoint method of ArcGIS 10.3 was used to classify the HAI into five levels: slight, weak, moderate, high, and strong. During the period of 2005–2020, the distribution of HAI in the WRB did not change significantly, showing a spatially strong southeastern and weak northwestern feature (Figure 5). As shown in Figure 6a, there is considerable spatial variation in the HAI levels in the WRB, taking the HAI level in 2005 as an example. The slight HAI occupied 52.66%, of which the Shaanxi section accounted for the most (26.66%), followed by the Gansu section (23.17%); the Ningxia section accounted for the least (2.83%). The proportion of the three sections in the weak HAI changed, with the proportion of the Gansu section accounting for 60.32% of the entire weak HAI, the Shaanxi section accounting for 29.81%, and the Ningxia section accounting for 9.89%. The proportion of the Ningxia section in moderate HAI was very small at 0.56%, and the proportion of the Shaanxi section returned to the first position with 86.85%. The proportion of the Gansu section was 12.58%. The HAI in the Ningxia section of the WRB contained only weak, mild, and moderate HAI. The high level of HAI accounted for 1.90% of the whole basin, of which the Shaanxi section had an absolute advantage with 91.80%, followed by the Gansu section with 8.44%. The strong HAI in the WRB had a very small percentage of 0.58%, of which the Shaanxi section accounted for 98.93%. In terms of changes in different HAI levels from 2005 to 2020, the slight and weak HAIs covered most of the study area, mainly in the northern and western parts of the WRB, accounting for more than 80% of the total basin, as shown in Figure 5. The Ningxia section was basically in the stage of weak HAI. The high and strong areas were mainly distributed in the Shaanxi section of the WRB, accounting for 3% of the total basin. The degree of slight HAI decreased significantly, from 52.66% in 2005 to 49.44% in 2020. The area percentage of weak and moderate HAI showed an increasing trend, with the percentage of weak HAI increasing from 32.03% in 2005 to 32.90% in 2020 and the percentage of moderate HAI increasing from 12.82% in 2005 to 14.61% in 2020. The percentages of high and strong HAI increased by 47.56% and decreased by 60.55%, respectively. In general, most of the regions were characterized by slight and weak HAI during the period of 2005–2020; the changes were not very significant, and the high and strong HAI was mainly concentrated in the Shaanxi section of the WRB.

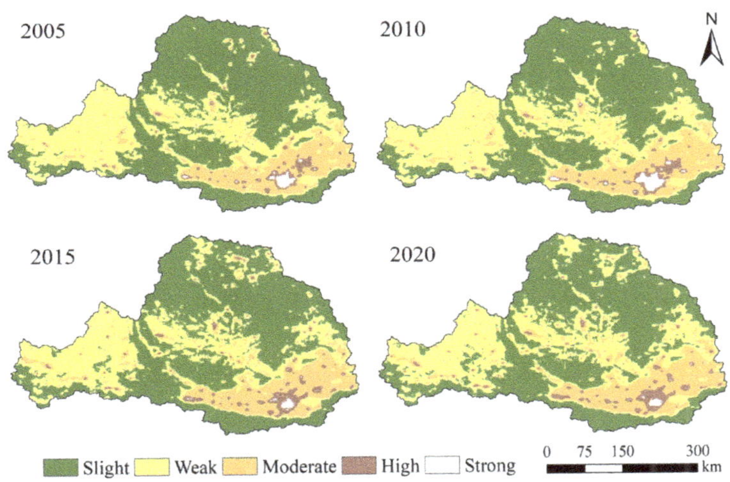

Figure 5. Spatial–temporal variation characteristics of HAI.

3.4. *Analysis of Anthropogenic Driving Factors of Soil Erosion Based on Geographic Detector Model*

The application results of the factor detector show that all five factors passed the 1% significance level test, and there are significant differences in the explanatory power of different factors on soil erosion in the WRB. Over time, the explanatory power of various factors on soil erosion in the watershed does not vary significantly, with POP and GDP having a higher explanation than PANDA and crop. Figure 7 shows the results of the interaction between factors, and it is found that the interaction results of any two factors present a dual factor or nonlinear enhancement. These results indicate that the spatial differences in soil erosion are caused by multiple influencing factors. The interaction between POP and cropland has the greatest explanatory power on soil erosion.

3.5. *Spatiotemporal Coupling Evolution between HAI and Soil Erosion*

The spatiotemporal coupling characteristics of HAI and soil erosion in the WRB were explored based on a four-quadrant model. As shown in Figure 8, there are obvious spatial differences between the coupling degree of HAI and soil erosion in the WRB from 2005 to

2020. The area with a coupling degree of poor is decreasing. Between 2005 and 2010, poor coupling degree was mainly distributed in the northern and western parts of the WRB, accounting for about 10%, and the area contracted to the north, reducing to about 3% of the study area from 2015 to 2020, which indicates a greater improvement in the ecosystem compared to 2005. The area with a coupling degree of general showed an increasing trend, with the area proportion increasing from 72.89% in 2005 to 78.44% in 2020, far exceeding the other three coupling degrees, indicating that the ecosystem is moving toward a stable trend. The area with a coupling degree of good is the smallest and is more stable, with the area being less than 0.5% in 2005–2020. The area with a coupling degree of excellent is mainly distributed in the southeast of the WRB, showing an overall increasing trend, with the area proportion increasing from 14.98% in 2005 to 17.36% in 2020, an increase of 15.89%.

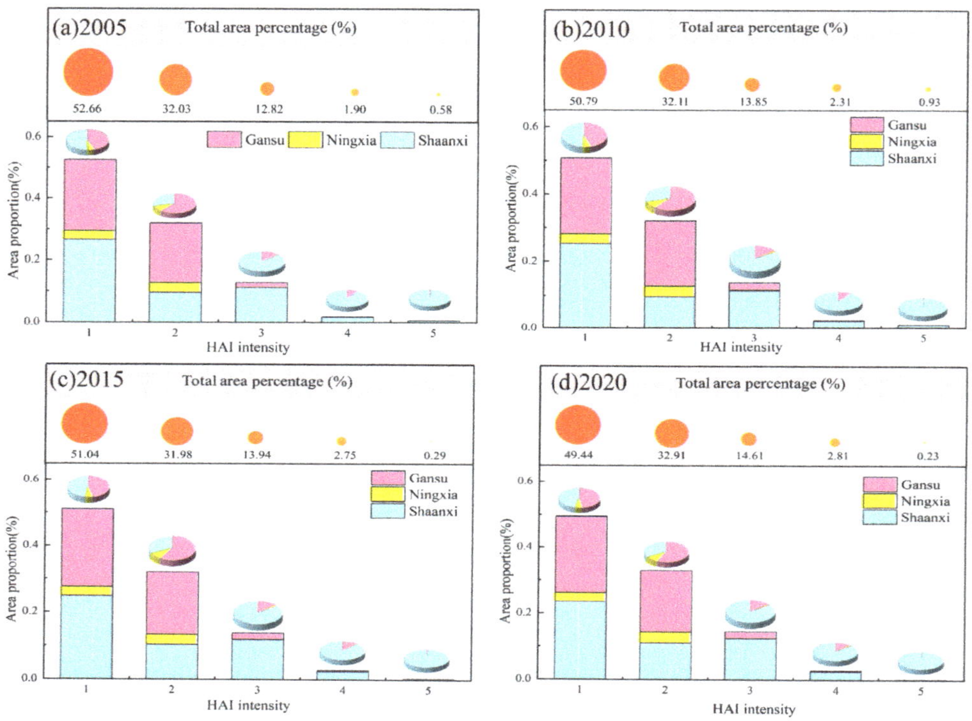

Figure 6. Changes in HAI in different provinces ((**a**) the change of HAI in the three provinces in 2005; (**b**) the change of HAI in the three provinces in 2010; (**c**) the change of HAI in the three provinces in 2015; (**d**) the change of HAI in the three provinces in 2020. Note: The columns represent the proportion of each province to the total HAI. The pie chart represents the proportion of each province in the same HAI, and the orange circle represents the proportion of each type of HAI).

Figure 9 illustrates the dynamics of the degree of coupling for different periods (2005–2010, 2010–2015, 2015–2020, and 2005–2020). From 2000 to 2010, the QFSI with the poor, general, good, and excellent coupling degrees were −0.228, 0.024, −0.075, and 0.536, respectively (Table 5). The outflow with a degree of coupling of poor is greater than the inflow, and its area tends to decrease, mainly shifting to the general coupling. The inflows are greater than the outflows in the excellent coupling degree, and their areas have a tendency to increase. The QFSI with good and general coupling degrees is close to 0, indicating that the change is not significant and is close to the equilibrium state of bidirectional conversion.

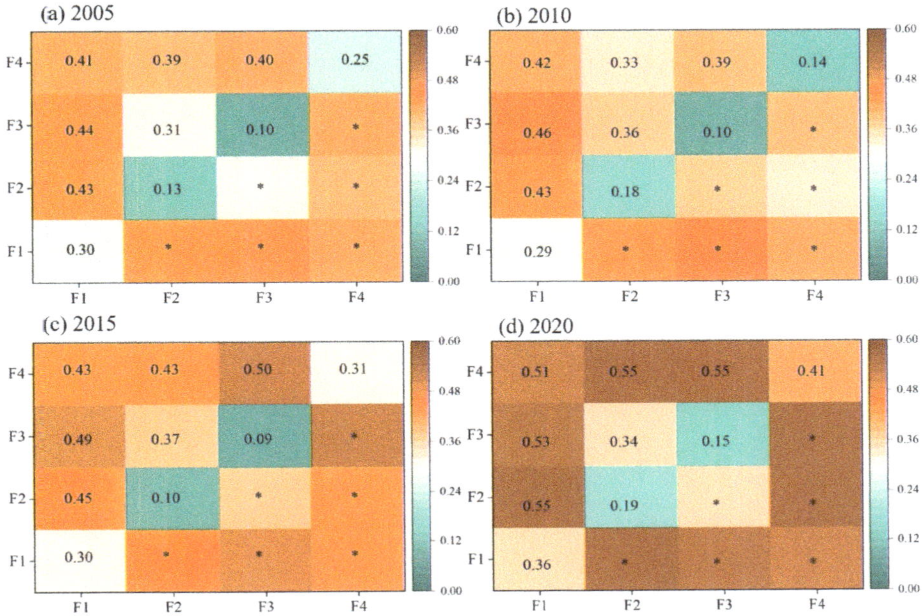

Figure 7. Variable correlation and interaction between 2005 and 2020 ((**a**) Driver correlations in 2005, (**b**) Driver correlations in 2010, (**c**) Driver correlations in 2015, (**d**) Driver correlations in 2020. Note: * represents p less than 0.05, F1 represents POP, F2 represents PANDA, F3 represents crop, and F4 represents GDP).

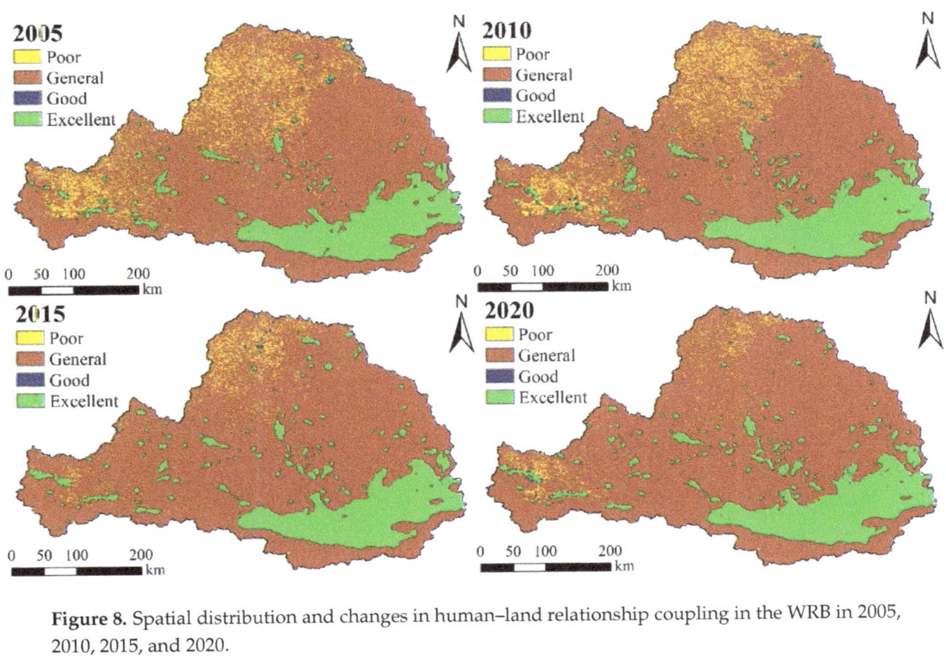

Figure 8. Spatial distribution and changes in human–land relationship coupling in the WRB in 2005, 2010, 2015, and 2020.

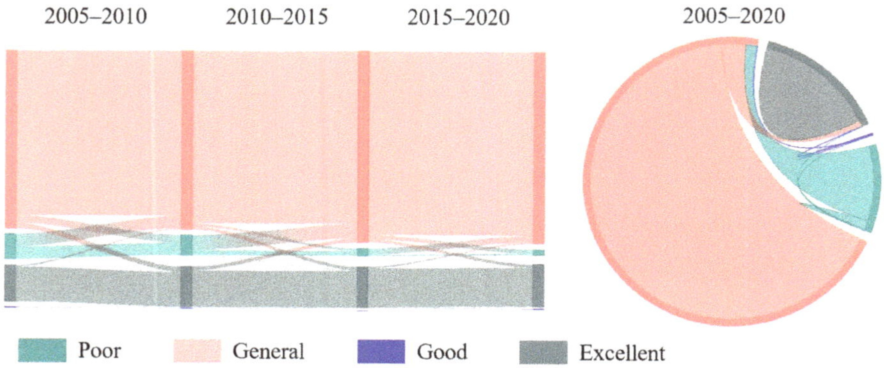

Figure 9. Dynamic changes in different coupling levels during different periods (2005–2010, 2010–2015, 2015–2020, and 2005–2020).

Table 5. QFSI with different coupling levels in different periods.

	Poor	General	Good	Excellent
2005–2010	−0.228	0.024	−0.075	0.536
2010–2015	−0.696	0.495	−0.578	0.040
2015–2020	−0.189	0.023	0.517	0.236
2005–2020	−0.746	0.396	−0.139	0.592

From 2010 to 2015, the QFSI with the poor, general, good, and excellent coupling degrees were −0.696, 0.495, −0.578, and 0.040, respectively (Table 5). Compared with 2005–2010, the area of the general coupling degree has increased significantly, while the areas of poor and good coupling degrees have decreased, among which the area with the poor coupling degree has decreased dramatically and the area with the general coupling degree is still at the highest percentage.

From 2015 to 2020, the QFSI with the poor, general, good, and excellent coupling degrees were −0.189, 0.023, 0.517, and 0.236, respectively (Table 5). Compared with 2010–2015, the area with the good, general, and excellent coupling degrees is increasing in varying degrees, indicating a steady increase in the effectiveness of ecological management in the WRB.

From 2005 to 2020, the QFSI with the good coupling degree is −0.139, indicating that the good human–land relationship is basically in a two-way transition equilibrium state. The area with the poor coupling degree where the relationship between humans and land is poor decreases a lot, and the QFSI is −0.746. The poor coupling degree mainly shifts to the general coupling degree. The general coupling degree has a partial flow into the fourth quadrant, indicating that the ecological environment has partially shifted to an excellent human–land relationship. Overall, it seems that the ecological environment of the WRB has gradually improved.

4. Discussion

The RUSLE method was used to quantify soil erosion in the WRB, and the temporal and spatial characteristics of soil erosion changes from 2005 to 2020 were discussed. In terms of space, the northern and western parts of the WRB are areas of relatively severe soil erosion. This heterogeneous spatial distribution is largely due to the special geographical environment, where the loess is deep and loose and prone to erosion [17,33]. The south-eastern area of the WRB is more economically developed, and soil erosion is relatively low because socio-economic development has promoted awareness of soil erosion and conscious changes in production and lifestyle [34]. In terms of time, the average soil erosion modulus in the WRB has decreased over time, indicating a change from relatively severe to weak soil erosion. The results of Sun et al. [35] are consistent with our findings, demonstrating that soil erosion in the Loess Plateau is decreasing at a rate of more than 1 $t \cdot km^{-2} \cdot a^{-1}$. Furthermore, the stability rate of soil erosion in the WRB was calculated, and the results showed that the stability rate of severe erosion in the WRB from 2005 to 2020 showed a downward trend. Among them, the transformation rate of severe erosion in the Shaanxi and Gansu sections was greater than 85%, and the transformation rate of severe erosion in the Ningxia section was as high as 96.28%. This may be attributed to the implementation of ecological protection projects, such as returning farmland to forests and grassland, and the GCHP project, which increased vegetation cover and improved the ecological environment of the WRB [24,36].

In addition, this study also found that the interaction between any two factors on soil erosion is greater than that of one factor and mainly exhibits a nonlinear enhancement effect, which is consistent with the findings of Xu et al. [23]. Soil erosion is affected by multiple factors. This study uses the geographical detector method to determine the driving factors that affect regional soil erosion changes. The results show that POP and crop have a greater impact than other factors and are dominant factors. Due to the implementation of ecological protection measures such as returning farmland to forests and grassland and GCHP since 1996, the overall ecosystem of the Loess Plateau has been restored and improved. For example, the forest cover, grass cover, and habitat quality have shown an increasing trend due to the implementation of the Grain for Green project [21]. Based on experiments, Xu et al. [14] came to the conclusion that GLC projects can restore and mitigate soil erosion while maintaining water and soil. Human activities are mainly concentrated in production and living activities such as the development and utilization of natural ecological resources; therefore, population density is also an important factor affecting soil erosion [37,38].

In this study, we analyzed the HAI in the WRB during the period of 2005–2020. Temporally, there is an upward trend in the percentage of area with weak, moderate, and high HAI. This indicates that the watershed ecosystem was effectively affected by human activities from 2005 to 2020, which is consistent with previous studies [39,40]. Xu and Xu [41] concluded that from 1992 to 2008, the number of county-level units included in the intermediate type area of the Loess Plateau increased substantially, and the mean value showed an overall upward trend. Spatially, the spatial distribution of HAI in the WRB shows strong southeast and weak northwest characteristics, which are related to urban population density. Human activities are mainly concentrated in production and living activities such as the development and utilization of natural ecological resources [36,37], so the higher values of HAI are mainly distributed in the southeastern part of the WRB. The northwestern region is less densely populated and has less space and fewer resources needed for human survival, so the Ningxia section as a whole is basically in a weak stage of HAI. This is consistent with the findings of previous studies on human activities in the WRB [17,21].

The relationship between HAI and soil erosion is dynamic. Our results show that about 97% of the region is distributed in the good, general, and excellent coupling degrees. Only 3% of the region is distributed in the poor coupling degree. This implies that the ecological quality of the WRB is generally good, and the coupling of HAI and soil erosion is good. The implementation of returning farmland to forests has improved the overall ecosystem service quality of the Loess Plateau [42,43]. In addition, Xu et al. [23] concluded based on the InVEST model that the soil conservation service capacity of the Jing River Basin improved after ecological conservation measures, such as GCHP. The proportion of good coupled regions gradually increased from 2005 to 2020. This may be related to human activities and climate change during the study period, consistent with previous studies [44]. In the 19th century, soil erosion rates on the Loess Plateau increased due to disturbances from human activities and interannual climate change [45]. However, with regard to the QFSI of the human–land relationship, the area of poor human–land relationship decreased between 2005 and 2020, with a QFSI of −0.746, in which there is mostly a flow to a general human–land relationship. Some of the general human–land relationship flows into the excellent human–land relationship flow. These basic characteristics of this study are consistent with the research results of Tan et al. [1] that the human–land relationship is constantly improving.

Identifying the impact of human activities on soil erosion can support regional soil and water conservation management and healthy development. It will be a long-term and gradual process to improve the coupling relationship between human activities and soil erosion, and the two will reach a good coupling state in order to achieve the goal of harmonizing the relationship between humans and the environment. Therefore, future policies should follow the principle of adapting measures to local conditions. For areas with strong human activities and high soil erosion, it is necessary to rationally allocate various land resources to avoid excessive reclamation and excessive use of land resources. Water and soil conservation measures such as dams and terraces should be implemented in the area. In areas with low human activity intensity and high soil erosion, environmental management and resource protection should be strengthened to enhance regional ecological resistance. For degraded land, vegetation restoration techniques can be used, such as artificial planting or natural vegetation growth to restore land vegetation. Shelter belts can also be established to reduce soil erosion. In some cases, there may be hydrologic factors unfavorable to afforestation in some areas. There is a practical need for a rational allocation of ecological conservation measures in the context of local economic development and urban planning. Due to the relationship between regional economy and soil erosion, it is possible to appropriately promote the development of cultural and tourism industries, increase the income of local residents, enhance their awareness of soil and water conservation, and reasonably develop land resources.

The quantification of HAI using the entropy weighting method has some limitations, which cannot take into account the cross-sectional effects between indicators [43]. The habitat of the WRB may face more anthropogenic influences than we have demonstrated, with studies suggesting that anthropogenic factors such as traffic volume and grazing intensity both impact the ecology [46,47]. Unfortunately, these factors must be excluded due to discontinuous inventory records. The above problems can lead to bias in the calculation of HAI [48]. Therefore, future work will comprehensively consider the impact of other factors on HAI, refine socio-economic indicators, and further explore the impact and contribution of regional industries to soil and water conservation. At the same time, we will improve the model to simulate the dynamic feedback process between human activities and soil erosion.

5. Conclusions

Anthropogenic disturbances to the ecosystem of the WRB will affect the spatial variation in soil erosion. Five types of landmark human activity factors were used to quantify the HAI from 2005 to 2020, and the spatiotemporal trends of HAI were explored from regional and provincial perspectives. Subsequently, the spatial–temporal variations of soil erosion in the WRB from 2005 to 2020 were quantified based on the RUSLE method, and the driving factors were analyzed using a geographical detector. Finally, based on the four-quadrant model, the coupling relationship between HAI and soil erosion was evaluated to identify areas with a low human–land relationship. The key findings can be summarized as follows:

(1) During the period of 2005–2020, the trend of soil erosion in the WRB shifted from high erosion levels to low erosion levels. Over the 15-year period, the erosion areas of moderate, high, strong, and severe decreased by 9.88%, 35.89%, 45.17%, and 70.05%, respectively. The areas of slight and weak erosion increased by 29.41% and 25.93%, respectively.

(2) HAI shows spatially strong southeast and weak northwest characteristics. The slight and weak HAI account for 80% in the northwest of the WRB, and the high HAI areas are mainly concentrated in the Shaanxi section.

(3) The overall human–land relationship in the WRB was generally stable and had a great improvement trend from 2005 to 2020. The proportion of general area in terms of the human–land relationship increased from 72.89% in 2005 to 78.44% in 2020. The distribution range of poor in the human–land relationship gradually shrank to about 3% of the study area.

Changes in regional economic and social factors, especially due to the implementation of ecological protection projects, may be the main reason for the improvement of human–land relationships across the region. The results are valuable for studying changes in human–land relations in the WRB and provide effective references for future ecosystem management.

Author Contributions: Conceptualization, Z.Z. and A.H.; methodology, Z.Z. and Q.L.; software, Z.Z. and Q.L.; validation, J.P., A.E. and A.H.; investigation, Z.Z. and Q.L.; data curation, Q.L. and M.E.-S.A.; writing—original draft preparation, Z.Z.; writing—review and editing, M.S.D.A.-H. All authors have read and agreed to the published version of the manuscript.

Funding: This research was funded by the National Natural Science Foundation of China (Grant No. 42261144749, 41790444 and 41877232); the Key Laboratory Open Project Fund of State Key Laboratory of Loess and Quaternary Geology, Institute of Earth Environment, CAS (Grant No. SKLLQG1909), Strategic Priority Program of the Chinese Academy of Sciences (No. XDA20030302); and Ningxia Natural Science Foundation of China (No. 2020AAC03476).

Institutional Review Board Statement: Research that does not involve humans or animals.

Informed Consent Statement: Research that does not involve humans or animals.

Data Availability Statement: Data will be made available on request.

Acknowledgments: The authors would like to extend their sincere appreciation to the National Natural Science Foundation of China (Grant No. 42261144749, 41790444 and 41877232); the Key Laboratory Open Project Fund of State Key Laboratory of Loess and Quaternary Geology, Institute of Earth Environment, CAS (Grant No. SKLLQG1909), Strategic Priority Program of the Chinese Academy of Sciences (No. XDA20030302); and Ningxia Natural Science Foundation of China (No. 2020AAC03476).

Conflicts of Interest: The authors declare that they have no known competing financial interests or personal relationships that could have appeared to influence the work reported in this paper.

References

1. Tan, S.; Liu, Q.; Han, S. Spatial-temporal evolution of coupling relationship between land development intensity and resources environment carrying capacity in China. *J. Environ. Manag.* **2022**, *301*, 113778. [CrossRef]
2. Barhagh, S.E.; Zarghami, M.; Ghale, Y.A.G.; Shahbazbegian, M.R. System dynamics to assess the effectiveness of restoration scenarios for the Urmia Lake: A prey-predator approach for the human-environment uncertain interactions. *J. Hydrol.* **2021**, *593*, 125891. [CrossRef]
3. Zhang, S.; Fan, W.; Li, Y.; Yi, Y. The influence of changes in land use and landscape patterns on soil erosion in a watershed. *Sci. Total Environ.* **2017**, *574*, 34–45. [CrossRef]
4. Singh, R.; Tiwari, K.N.; Mal, B.C. Hydrological studies for small watershed in India using the ANSWERS model. *J. Hydrol.* **2006**, *318*, 184–199. [CrossRef]
5. Cunha, J.E.F.; Bravo, J.V.M. Effects of environmental protection policies on fragile areas of a watershed occupied by agriculture in the Brazilian Cerrado. *J. Environ. Manag.* **2022**, *319*, 115695. [CrossRef] [PubMed]
6. Huo, A.; Yang, L.; Luo, P.; Cheng, Y.; Peng, J.; Nover, D. Influence of landfill and land use scenario on runoff, evapotranspiration, and sediment yield over the Chinese Loess Plateau. *Ecol. Indic.* **2021**, *121*, 107208. [CrossRef]
7. Huo, A.; Yang, L.; Peng, J.; Cheng, Y.; Jiang, C. Spatial characteristics of the rainfall induced landslides in the Chinese Loess Plateau. *Hum. Ecol. Risk Assess.* **2020**, *26*, 2462–2477. [CrossRef]
8. Luo, P.; Luo, M.; Li, F.; Qi, X.; Huo, A.; Wang, Z.; He, B.; Takara, K.; Nover, D.; Wang, Y. Urban flood numerical simulation: Research, methods and future perspectives. *Environ. Model. Softw.* **2022**, *156*, 105478. [CrossRef]
9. Schiettecatte, W.; D´hondt, L.; Cornelis, W.; Acosta, M.; Leal, Z.; Lauwers, N.; Almoza, Y.; Alonso, G.; Díaz, J.; Ruíz, M.; et al. Influence of landuse on soil erosion risk in the Cuyaguateje watershed (Cuba). *CATENA* **2008**, *74*, 1–12. [CrossRef]
10. Foucher, A.; Evrard, O.; Cerdan, O.; Chabert, C.; Lefèvre, I.; Vandromme, R.; Salvador-Blanes, S. Deciphering human and climatic controls on soil erosion in intensively cultivated landscapes after 1950 (Loire Valley, France). *Anthropocene* **2021**, *34*, 100287. [CrossRef]

11. Mhaske, S.N.; Pathak, K.; Dash, S.S.; Nayak, D.B. Assessment and management of soil erosion in the hilltop mining dominated catchment using GIS integrated RUSLE model. *J. Environ. Manag.* **2021**, *294*, 112987. [CrossRef] [PubMed]
12. Sreedevi, S.; Eldho, T.I.; Jayasankar, T. Physically-based distributed modelling of the hydrology and soil erosion under changes in landuse and climate of a humid tropical river basin. *CATENA* **2022**, *217*, 106427. [CrossRef]
13. Zhang, Z.; Xu, W.; Li, L.; Huang, J.; Deng, L.; Wang, Q. Effects of temporal conservation measures on water erosion processes of disturbed soil accumulation in construction projects. *J. Clean. Prod.* **2021**, *319*, 128612. [CrossRef]
14. Xu, Q.; Chen, W.; Zhao, K.; Zhou, X.; Du, P.; Guo, C.; Ju, Y.; Pu, C. Effects of land-use management on soil erosion: A case study in a typical watershed of the hilly and gully region on the Loess Plateau of China. *CATENA* **2021**, *206*, 105551. [CrossRef]
15. Li, W.; Geng, J.; Bao, J.; Lin, W.; Wu, Z.; Fan, S. Analysis of Spatial and Temporal Variations in Ecosystem Service Functions and Drivers in Anxi County Based on the InVEST Model. *Sustainability* **2023**, *15*, 10153. [CrossRef]
16. Sun, Y.; Liu, S.; Shi, F.; An, Y.; Li, M.; Liu, Y. Spatio-temporal variations and coupling of human activity intensity and ecosystem services based on the four-quadrant model on the Qinghai-Tibet Plateau. *Sci. Total Environ.* **2020**, *743*, 140721. [CrossRef]
17. Jia, L.; Yao, S.; Deng, Y.; Li, Y.; Hou, G.; Gong, Z. Temporal and Spatial Characteristics of Soil Erosion Risk in Weihe River Basin and Its Geographical Exploration. *J. Ecol. Rural Environ.* **2021**, *37*, 305–314. [CrossRef]
18. Li, N.; Zhang, Y.; Wang, T.; Li, J.; Yang, J.; Luo, M. Have anthropogenic factors mitigated or intensified soil erosion over the past three decades in South China? *J. Environ. Manag.* **2022**, *302*, 114093. [CrossRef]
19. Wang, J.; Wang, Z.; Li, K.; Li, C.; Wen, F.; Shi, Z. Factors affecting phase change in coupling coordination between population, crop yield, and soil erosion in China's 281 cities. *Land Use Policy* **2023**, *132*, 106761. [CrossRef]
20. Liu, Y.; Zhao, Y.; Xu, G. Correlation between ecosystem services value and human activity intensity based on the four-quadrant model: A case study in the International Tourism and Culture Demonstration Area, the South Anhui Province. *Acta Ecol. Sin.* **2022**, *42*, 5200–5210.
21. Xu, X.; Zhao, X.; Song, X. Impacts of the returning farmland to forest (grassland) project on ecosystem services in the Weihe River Basin, China. *Chin. J. Appl. Ecol.* **2021**, *32*, 3893–3904. [CrossRef]
22. Hou, K.; Tao, W.; Wang, L.; Li, X. Study on hierarchical transformation mechanisms of regional ecological vulnerability and its applicability. *Ecol. Indic.* **2020**, *114*, 106343. [CrossRef]
23. Xu, C.; Jiang, Y.; Su, Z.; Liu, Y.; Lyu, J. Assessing the impacts of Grain-for-Green Programme on ecosystem services in Jinghe River basin, China. *Ecol. Indic.* **2022**, *137*, 108757. [CrossRef]
24. Sun, C.; Liu, Y.; Song, H.; Li, Q.; Cai, Q.; Wang, L.; Fang, C.; Liu, R. Tree-ring evidence of the impacts of climate change and agricultural cultivation on vegetation coverage in the upper reaches of the Weihe River, northwest China. *Sci. Total Environ.* **2020**, *707*, 136160. [CrossRef]
25. Park, S.; Oh, C.; Jeon, S.; Jung, H.; Choi, C. Soil erosion risk in Korean watersheds, assessed using the revised universal soil loss equation. *J. Hydrol.* **2011**, *399*, 263–273. [CrossRef]
26. Zhang, W.; Fu, J. Rainfall Erosivity Estimation Under Different Rainfall Amount. *Resour. Sci.* **2003**, *25*, 35–41.
27. Williams, J.R.; Arnold, J.G. A system of erosion—Sediment yield models. *Soil Technol.* **1997**, *11*, 43–55. [CrossRef]
28. Cao, S.; Ouyang, M.; Zhou, W. Quantitative evaluation of soil erosion in Ningxiang City based on GIS and USLE. *J. China Agric. Univ.* **2018**, *23*, 149–157.
29. Almagro, A.; Thomé, T.; Colman, C.; Pereira, R.; Junior, J.; Rodrigues, D.; Oliveira, P. Improving cover and management factor (C-factor) estimation using remote sensing approaches for tropical regions. *Int. Soil Water Conserv. Res.* **2019**, *7*, 325–334. [CrossRef]
30. Zhong, D. *Study on Soil Erosion in Loess Hill and Gully Area on USLE Model: A Case Study of Wuqi County*; Northwest A&F University: Xianyang, China, 2012.
31. Wang, Q.; Wu, C.; Sun, Y. Evaluating corporate social responsibility of airlines using entropy weight and grey relation analysis. *J. Air Transp. Manag.* **2015**, *42*, 55–62. [CrossRef]
32. Ma, S.; Xie, F.; Ding, C.; Zhang, H. Spatio-temporal change of landscape ecological quality and influencing factors based on four-quadrant model in overlapped area of cropland and coal production. *Trans. Chin. Soc. Agric. Eng.* **2020**, *36*, 259–268.
33. Cheng, Y.; Huo, A.; Zhao, Z.; Peng, J. Analysis of loess fracture on slope stability based on centrifugal model tests. *Bull. Eng. Geol. Environ.* **2021**, *80*, 3647–3657. [CrossRef]
34. Dotterweich, M. The history of human-induced soil erosion: Geomorphic legacies, early descriptions and research, and the development of soil conservation—A global synopsis. *Geomorphology* **2013**, *201*, 1–34. [CrossRef]
35. Sun, W.; Shao, Q.; Liu, J. Soil erosion and its response to the changes of precipitation and vegetation cover on the Loess Plateau. *J. Geogr. Sci.* **2013**, *23*, 1091–1106. [CrossRef]
36. Zhu, Y.; Luo, P.; Zhang, S.; Sun, B. Spatiotemporal Analysis of Hydrological Variations and Their Impacts on Vegetation in Semiarid Areas from Multiple Satellite Data. *Remote Sens.* **2020**, *12*, 4177. [CrossRef]
37. Zhang, F.; Wang, Y.; Ma, X.; Wang, Y.; Yang, G.; Zhu, L. Evaluation of resources and environmental carrying capacity of 36 large cities in China based on a support-pressure coupling mechanism. *Sci. Total Environ.* **2019**, *688*, 838–854. [CrossRef] [PubMed]
38. Zhang, P.; Yang, D.; Qin, M.; Jing, W. Spatial heterogeneity analysis and driving forces exploring of built-up land development intensity in Chinese prefecture-level cities and implications for future Urban Land intensive use. *Land Use Policy* **2020**, *99*, 104958. [CrossRef]

39. Du, J.; Shi, C. Effects of climatic factors and human activities on runoff of the Weihe River in recent decades. *Quat. Int.* **2012**, *282*, 58–65. [CrossRef]
40. Zhang, T.; Su, X.; Zhang, G.; Wu, H.; Wang, G.; Chu, J. Evaluation of the impacts of human activities on propagation from meteorological drought to hydrological drought in the Weihe River Basin, China. *Sci. Total Environ.* **2022**, *819*, 153030. [CrossRef]
41. Xu, X.; Xu, Y. Analysis of spatial-temporal variation of human activity intensity in Loess Plateau region. *Geogr. Res.* **2017**, *36*, 661–672.
42. Sun, D.; Li, H.; Dawson, R.; Tang, C.; Li, X. Characteristics of Steep Cultivated Land and the Impact of the Grain-for-Green Policy in China. *Pedosphere* **2006**, *16*, 215–223. [CrossRef]
43. Wang, Y.; Zhao, J.; Fu, J.; Wei, W. Effects of the Grain for Green Program on the water ecosystem services in an arid area of China—Using the Shiyang River Basin as an example. *Ecol. Indic.* **2019**, *104*, 659–668. [CrossRef]
44. Zhao, A.; Zhu, X.; Liu, X.; Pan, Y.; Zuo, D. Impacts of land use change and climate variability on green and blue water resources in the Weihe River Basin of northwest China. *CATENA* **2016**, *137*, 318–327. [CrossRef]
45. Wang, X.; Wang, Z.; Xiao, J.; He, M.; Zhang, F.; Pan, Y.; Zhang, Y.; Jin, Z. Soil erosion fluxes on the central Chinese Loess Plateau during CE 1811 to 1996 and the roles of monsoon storms and human activities. *CATENA* **2021**, *200*, 105148. [CrossRef]
46. Holden, P.B.; Ziervogel, G.; Hoffman, M.T.; New, M.G. Transition from subsistence grazing to nature-based recreation: A nuanced view of land abandonment in a mountain social-ecological system, southwestern Cape, South Africa. *Land Use Policy* **2021**, *105*, 105429. [CrossRef]
47. Wang, X.; Birch, G.F.; Liu, E. Traffic emission dominates the spatial variations of metal contamination and ecological-health risks in urban park soil. *Chemosphere* **2022**, *297*, 134155. [CrossRef]
48. Burkhard, B.; Müller, A.; Müller, F.; Grescho, V.; Anh, Q.; Arida, G.; Bustamante, J.; Chien, H.; Heong, K.; Escalada, M.; et al. Land cover-based ecosystem service assessment of irrigated rice cropping systems in southeast Asia—An explorative study. *Ecosyst. Serv.* **2015**, *14*, 76–87. [CrossRef]

sustainability

MDPI

Article

The Study on Solving Large Pore Heat Transfer Simulation in Malan Loess Based on Volume Averaging Method Combined with CT Scan Images

Yangchun Lu [1,2], Ting Lu [1,2], Yudong Lu [1,2,*], Bo Wang [1,2], Guanghao Zeng [1,2] and Xu Zhang [1,2]

[1] School of Water and Environment, Chang'an University, Xi'an 710054, China
[2] Key Laboratory of Subsurface Hydrology and Ecological Effect in Arid Region of Ministry of Education, Chang'an University, Xi'an 710054, China
* Correspondence: luyudong@chd.edu.cn

Abstract: Malan loess is a wind-formed sediment in arid and semi-arid regions and is an important constituent of the Earth's critical zone. Therefore, the study of the relationship between microstructure and heat transfer in Malan loess is of great significance for the in-depth understanding of the heat transfer mechanism and the accurate prediction of the heat transfer properties of intact loess. In order to quantitatively characterize the heat transfer processes in the two-phase medium of solid particles and gas pores in the intact loess, this study used modern computed tomography to CT scan the Malan loess in Huan County, Gansu Province, the western part of the Loess Plateau, China and used the specific yield of the intact Malan loess as the parameter basis for extracting the threshold segmentation of the large pores in the scanned images for the three-dimensional reconstruction of the connected large pores. An experimental space for heat conduction of intact Malan loess was constructed, and the surface temperature of Malan loess was measured on the surface of the space with a thermal imager. The simulation of the heat conduction process was carried out using the solution program in AVIZO (2019) software using the volume averaging method combined with CT scanning to reconstruct the 3D pores. The experiments of heat conduction in the intact Malan loess showed that for a given external temperature pressure, the temperature decreases along the heat flow direction as a whole. The temperature of the pores in the normal plane along the heat flow direction is higher than the temperature of the solid skeleton. Abnormal temperature points were formed at the junction of the surface and internal pores of Maran loess, and the temperature of the jointed macropores was about 1 °C higher at the surface of the sample than that of the surrounding solid skeleton. Simulation of heat conduction in Malan loess showed that the heat transfer process in Malan loess was preferentially conducted along the large pores and then the heat was transferred to the surrounding Malan loess particle skeleton. The simulation results of heat conduction in Malan loess were in high agreement with the experimental results of heat conduction in Malan loess, which verifies the reliability of the calculated model.

Keywords: CT scan; intact Malan loess; volume averaging; heat transfer simulation; thermal imaging introduction

1. Introduction

Malan loess is a typical porous medium. Malan loess is located in the upper layer of the loess plateau in the vertical space, which is a penetrating area [1,2]. The spatial extent of Malan loess starts from the surface and extends through the loess layer to the bottom of groundwater, a critical zone for the survival of terrestrial life on the Loess Plateau, which is essential for sustaining life on the Loess Plateau [3,4]. Malan loess is located in the boundary area of the Earth's five major circles and is in a dynamic evolutionary process at all times. Water and gas flow in Malan loess, and all this is the result of the synergistic effect of biology, geology, and climate on Malan loess in the geological evolution process [5,6].

Malan loess is one of the most common materials in the Loess Plateau, the core part of the Earth's critical zone, and one of the most heterogeneous and complex areas in the critical zone [7,8]. Due to its relatively recent deposition, Malan loess is subject to less pressure from the overlying strata and is less dense compared to the loess of other loess layers. During the process of deposition, it was subjected to the action of rainfall, temperature, and plant roots. Therefore, a variety of pore structures have been formed during the evolution of Malan loess. The process of heat transfer in loess affects the hydro-ecological process of the whole loess area. Malan loess is closer to the surface and is greatly affected by climate change, so the process of heat transfer in Malan loess becomes very complicated [9].

Nowadays, in most loess soil physics, the study of heat transfer in loess soil is treated as a homogeneous object, and the temperature of one point in the soil is measured as the temperature of the whole soil body [10–13]. However, in nature, Malan loess is non-homogeneous and anisotropic, so the heat transfer in Malan loess can be different. Most studies have focused on determining heat transfer processes in soils through soil moisture temperature sensors, a method that involves burying a probe into the measured soil during the process of determining soil heat transfer, which changes the natural deposition state of the soil and uses the measured value of one probe as the change in temperature of the entire soil around that probe [14–17]. However, the real heat transfer process in nature in Malan loess is along the pore channels of Malan loess and the connection of the particle skeleton. It is a dynamic process rather than a constant value throughout the heat transfer interface of Malan loess. The industrial-grade thermal imager monitors temperature changes across the face in real time and can detect temperatures at various points on a face. This non-contact temperature measurement does not change the original structural characteristics of the Malan loess [18,19].

Heat transfer in porous and particulate media is a widely studied topic mainly in the chemical engineering systems and materials [20]. In general, the effect of solid particles on the thermal conductivity of porous media is not dominant, especially for those non-metallic porous materials [12,21–23]. The thermal conductivity in Malan loess porous media is very complex and there are many factors affecting the thermal conductivity process of Malan loess, including the physical properties of the solid particle skeleton of Malan loess, the pore size shape and distribution inside Malan loess, the type of fluid, the component morphology, and properties, etc. Pressure and temperature also have an effect on the thermal conductivity process of Malan loess [10–12,24]. In the analysis of thermal conductivity processes in porous media, two methods are usually combined, the first simplifies the physical model and the second introduces the effective thermal conductivity [25–28].

The pore structure of Malan loess is a hot issue in the study of pore heat transfer in Malan loess. With the application of photoelectric microscopy, image processing, and X-ray diffraction, the morphological characteristics of pores as well as particle skeleton in Malan loess can be extracted [9]. With the development of X-ray computed tomography (CT), it has made new progress in the study of the three-dimensional structure of loess by virtue of nondestructive, quantitative, and repeatable observation [29–33]. The acquisition of three-dimensional fine view pore channels in loess is mainly conducted by industrial CT scanning and then by threshold division to extract the pore channels in loess, but CT scanning is a process by X-ray diffraction, which can only capture the morphology of pore and particle skeleton but cannot capture the process of heat transfer in loess.

The process of studying heat transfer in the porous media of Malan loess requires linking the macroscopic, fine, and microscopic aspects of the multiphase pore media of Malan loess [34–36]. Compared with the classical theoretical coupling method and mixture theory, the volume averaging method not only realizes the coupling of heat flow processes in multiphase porous media studied within a unified hydrodynamic framework but also allows the derivation of effective local volume averaging equations with pore-scale information. The volume averaging method simultaneously performs volume averaging of the microscopic equilibrium equation and the microscopic intrinsic structure equation to obtain the macroscopic equilibrium equation and the macroscopic intrinsic structure equation [37–39].

With the development of 3D visualization software, the extraction of pore and particle skeletons and their morphological parameters in porous media based on CT scans has become more convenient. Based on the solution of the closure problem, a simulation package for heat transport in porous media is developed using AVIZO software combined with CT scan images using the volume averaging method. Using this solution procedure, the images obtained from CT scans can be analyzed and heat transport simulations of porous media can be performed.

In this study, we used CT scanning technology to scan the intact Malan loess sample in Huan County, western part of the Loess Plateau, China, and constructed a heat transport space for the intact Malan loess sample, and simulated the heat transport of the intact Malan loess sample in a two-phase medium (solid and gas) in the vertical direction by controlling the temperature boundary at the bottom of the Malan loess sample and monitoring the temperature with a thermal imager at the top of the sample. The process of heat transport in the two-phase medium (solid and gas) was simulated in the vertical direction. The volume averaging method combined with the 3D pore space reconstructed by CT scan was used to simulate the heat transport process in the intact Malan loess sample. The study of heat transfer in the microstructure of intact Malan loess can be important in helping to understand the thermodynamic characteristics of soils as well as in modeling and predicting water movement, and soil temperature distribution.

2. Study Area and Dataset

2.1. Collection and Preparation of Malan Loess Samples

According to regional geological data, in Northwestern China, the loess area in the middle of the Loess Plateau, the thickness of Malan loess on West Mountain of Huanxian County, Qingyang City, Gansu Province, is between 0–25 m, and the deposition thickness of Lishi loess is between 5–80 m. In the loess layer, there is a red ancient soil layer interspersed, and below the loess layer is a sandy cobble and gravel layer with a thickness greater than 200 m. The whole area has a complete sequence of loess development. The sampling point was arranged on the West Hill of Huanxian County, with the sampling coordinates of 107.284294° E, 36.562462° N, and 1462 m above sea level, and the intact Malan loess was taken at 1.5 m from the top of the surface in the Malan loess deposition area on the top of the West Hill [40–42], as shown in Figure 1. A complete bulk sample of intact Malan loess was first taken to ensure that the original deposition direction was unchanged, and then the bulk sample was cut into a cylinder with a diameter of 10 cm and a height of 10 cm. The normal direction of the cylinder diameter was the same as the deposition direction. The samples were carefully packed with foam and brought to the laboratory, and then dried in an oven at 105 °C for 10 h. The weight of the samples was measured to be 956 g. The dry density of the samples was 1.219 g/cm^3. The excess soil samples were cut out during the sample preparation and the basic physical properties of the loess samples were determined by a combined Malvern Mastersizer 2000 laser analyzer and a photoelectric type liquid limit plastic limit meter, the permeability coefficient of soil samples was measured with TST-55 variable head permeameter, and the results are shown in Table 1.

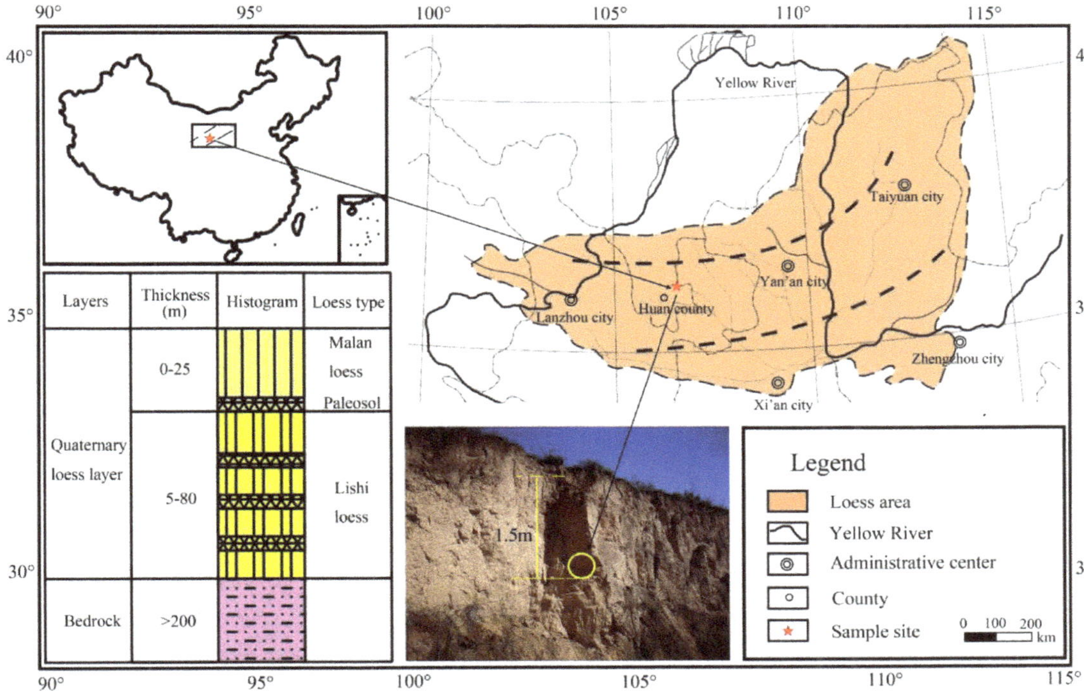

Figure 1. Location map of sampling points. (The yellow circle is the sampling position on the field section of Malan Loess).

Table 1. Basic physical properties of Malan loess samples.

Dry Density (g/cm^3)	Void Ratio	Clay (%) ($d < 5$ μm)	Silt (%) ($5 < d < 50$ μm)	Sand (%) ($d > 50$ μm)	Liquid Limit	Plastic Limit	Plastic Index	Permeability Coefficient K (cm/s)
1.219	1.18	9.46	61.07	29.74	11.5%	8%	11.83%	1.38×10^{-4}

2.2. CT Scan

This CT scan experiment was conducted at the Highway Laboratory of Chang'an University, Xi'an, Shaanxi Province. CT scanning is a non-destructive testing technique that can be used to obtain the internal structure of a sample and has a wide range of applications in the study of porous media. In conducting CT scanning experiments on Malan loess, X-rays passing through the intact Malan loess sample were attenuated and the reflected signals were accepted by the ray detector and converted into electrical signals and then into a series of gray-scale image slices. In order to obtain the ideal CT scan slice, the final settings were 180 KV for the scan voltage, 0.35 mA for the current, 300 ms (integration time), 1.40 (magnification), Y.XRD1620 for the detector type, and 204.80 for the detector length. By setting this set of scanning parameters, finally, clear grayscale image slices of this scanned sample were obtained. There were 712 slices in the vertical direction with a resolution of 140.349 μm, a magnification of 1.41, and a pixel size of 1024 × 1024. The scanned image was imported into the software, and the image is first spatially calibrated according to the grayscale image obtained from the scan. The slice overlay image generated after calibration is shown in Figure 2a, and the original 3D model of the Malan loess sample is reconstructed as shown in Figure 2b.

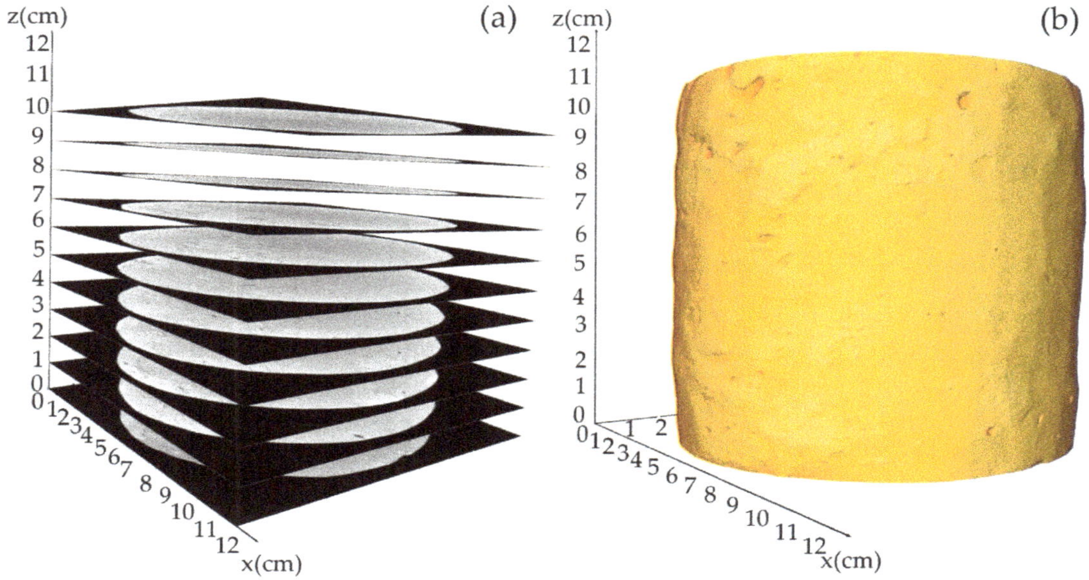

Figure 2. (**a**) Overlay of CT scan sections of intact Malan loess; (**b**) 3D model of reconstructed Malan loess sample.

3. Methods

3.1. Intact Loess Heat Conduction Experiment

The process of heat transfer in porous media is very complex, and in order to simplify the computational model, some boundary conditions must be set in conducting experiments on thermal conductivity of intact Malan loess samples. The present experiments on heat conduction of intact Malan loess are conducted physically with a homemade observation device for preferential channeling of heat in large pores of intact loess samples. By using the combination of temperature control knob, silica nanoaerogel mat, silica heating plate, and foam adhesive, a space for heat transfer action is provided for the intact loess so that the loess specimen can be heated uniformly at the bottom and the skeletal structure of the loess can be kept unchanged without loss of heat in the lateral direction. The heat enters from the bottom of the intact loess specimen and is conducted upward. The temperature of the silica gel heating plate can be controlled by the temperature control knob, and the heat conduction experiment with different input temperature changes can be completed. The infrared thermal imager on the surface of the intact loess specimen can monitor the heat distribution on the surface of the intact Malan loess sample and can continuously identify the preferred heat transfer channels of the large pores in the intact loess. The thermal imager can monitor the abnormal temperature points on the surface of the intact Malan loess sample.

3.1.1. Experimental Design and Experimental Setup

The present two-phase (air, particle skeleton) media thermal conductivity experimental setup for the intact Malan loess sample is shown in Figure 3. Among them, 1 is the power supply, 2 is the temperature control knob, 3 is the silica gel heating plate, 4 is the PVC fixing tube, 5 is the foam adhesive, 6 is the original cylindrical Malan loess, 7 is the infrared thermal imaging camera, 8 is the bracket, and 9 is the computer.

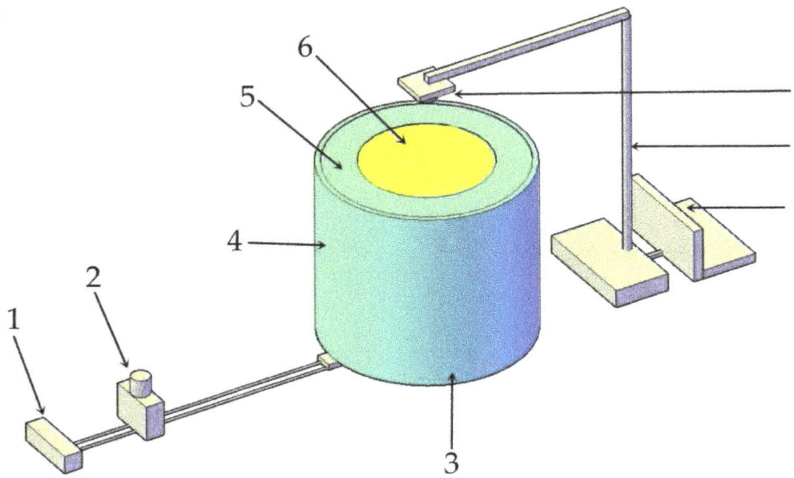

Figure 3. Schematic diagram of thermal imaging experiment: 1 is the power supply, 2 is the temperature control knob, 3 is the silica gel heating plate, 4 is the PVC fixing tube, 5 is the foam adhesive, 6 is the original cylindrical Malan loess, 7 is the infrared thermal imaging camera, 8 is the bracket, and 9 is the computer.

The power supply voltage was 220 V, where the adjustment range of the temperature control knob is from 30 °C to 150 °C, and the adjustment accuracy was ±2 °C. The silicone heating plate connected to the temperature control knob is a round plate with a diameter of 10 cm and a thickness of 0.8 cm. The temperature measurement device uses Hikvision H21PRO high-precision industrial pyrometer for thermal imaging temperature measurement. The resolution of the detector was 256 × 192, the temperature measurement range was −20 °C–350 °C, the temperature measurement accuracy is ±0.2 °C, the minimum temperature measurement distance is 15 cm, and 19,200 temperatures can be detected simultaneously.

In order to disregard the lateral heat loss of the Malan loess sample during heat transfer, a lateral insulation device was required on the side of the Malan loess, as shown in Figure 3 (4, 5, and 6). A PVC pipe with an inner diameter of 16 cm and a wall thickness of 3 mm was used for the outer support wall of the insulation device, which could serve as a lateral support. Since the surface of the Malan loess sample is relatively rough, it is difficult for the general solid insulation material to achieve absolute insulation of the sidewall. This time, we used polyurethane foam, which can grow automatically and has excellent high-temperature resistance, and the temperature can reach 100 °C to 150 °C, which is less than the temperature designed for this experiment. The foam glue has high foaming times, good stability, small foam pore size, uniform and fine, which can effectively fill the pores between the fixed sidewall and the Malan loess sample. The specific operation steps were as follows: In the first step, a layer of 1 mm latex was applied to the side surface of the dry intact Malan loess sample after the CT scan experiment, and the sample was left to rest for 24 h until the glue completely adhered to the surface of the Malan loess. This is used to seal the pores on the side surface of the Malan loess and also to prevent the infiltration of the foam glue into the Malan loess sample. In the second step, put the original cylindrical Malan loess sample into the middle of the PVC support wall, and protect the upper top surface of the Malan loess sample with a 10 cm diameter piece of oil-based paper. Pour the mixed Styrofoam evenly along the glass rod into the space between the sample and the PVC support wall. Then the Styrofoam started to grow upwards along the gap and eventually stopped growing and curing. In the third step, wait until the foam adhesive was completely solidified, remove the oil-based protective paper on the top surface of the cylindrical sample, and then glue the silica gel heating plate on the bottom surface of the cylindrical Malan loess sample, and wrap the outer layer with silica nanoaerogel felt. The silica nanoaerogel felt has good heat insulation and thermal insulation effect, which can effectively isolate heat loss.

3.1.2. Thermal Imaging Experiments

The infrared thermal imager was installed directly above the sample, the temperature measurement distance was adjusted to 15 cm, and the material emissivity was adjusted to 0.92. The material emissivity was regulated according to the soil emissivity. Connect all the test devices as shown in Figure 3, then turn on the power, adjust the temperature control knob to 100 °C, check the temperature change at the connection to see if there is heat loss, check the thermal insulation effect of the device, and rest the sample for 12 h. Continue to adjust the temperature control knob to 80 °C, adjust the emissivity to 0.92 on the infrared thermal imager, and adjust the temperature measurement distance. Next, imaging is performed every minute. In order to capture the process of the occurrence of different temperature points, the imaging frequency was encrypted when the occurrence of different temperature points was detected for the first time.

3.1.3. Thermal Imaging Validation Experiments

A validation experiment was designed to verify whether this heat conduction process at the surface temperature variability of Malan loess is conducted by the particle-to-particle skeletal connections in Malan loess or due to the pore spaces in Malan loess. A series of straight pore channels were artificially created in the cylindrical primary Malan loess to simulate the pore channels in the Malan soil. The four holes have diameters of 0.5 mm, 1 mm, 1.5 mm, and 1.8 mm, each hole has a depth of 70 mm, the distance between each two holes in the transverse direction was 20 mm, and the distance between two holes of the same diameter in the longitudinal direction was 30 mm. The distribution of the artificial holes in the plane is shown in Figure 4a, and the perspective view of the artificial holes is shown in Figure 4b. The artificial holes did not penetrate the whole cylindrical sample, leaving 30 mm of intact Malan loess structure on the bottom surface, which allowed the heat to be transferred in the Malan loess sample for a distance first, avoiding the thermal imager to detect the temperature on the surface of the heating plate directly.

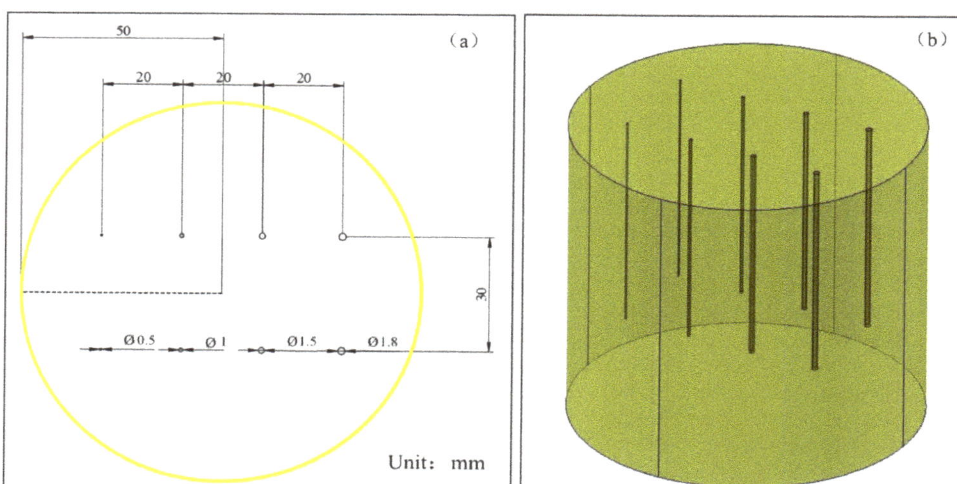

Figure 4. Schematic diagram of sample perforation. (**a**) The distribution of the artificial holes in the plane; (**b**) The perspective view of the artificial holes.

3.2. Heat Conduction Simulation of Intact Malan Loess

In order to study the process of solid–gas two-phase heat transfer in the porous media Malan loess, this section describes the application of multiscale volume average theory to Fourier's law for the heat transfer equation.

3.2.1. Fundamentals

Heat transfer is the ability of a material to conduct heat from a high-temperature region to a low-temperature region. In order to solve the process of heat transfer in a solid and gas phase medium of intact Malan loess at a given temperature–pressure condition, the heat conduction equation needs to be introduced. Under steady-state conditions, the heat conduction of a homogeneous material is described by Fourier's law [14–17].

$$\vec{\varphi} = -\lambda\vec{\nabla}T \tag{1}$$

The intact Malan loess is a porous medium material that is non-homogeneous. The thermal conductivity process in drying Malan loess samples contains solid particle skeleton thermal conductivity and pore skeleton thermal conductivity, and the thermal conductivity medium has two phases: solid and gas. The role of gas phase thermal conductivity at a given temperature and pressure cannot be neglected. The transient heat conduction of a given single-phase medium (α) phase is described by a partial differential equation [14].

$$\left(\rho c_p\right)_\alpha \frac{\partial T_\alpha}{\partial t} - \lambda_\alpha \nabla^2 T_\alpha = 0 \tag{2}$$

When the heat transfer state reaches stability, the equation to be solved is:

$$\lambda_\alpha \nabla^2 T_\alpha = 0 \tag{3}$$

In the physical experiments of thermal conductivity of two-phase media (gas and solid) of intact Malan loess cylindrical intact Malan loess samples a constant heat flow was applied on the bottom surface with a heating plate, on the top surface of the sample and air contact surface is ambient temperature, and the sample was surrounded by thermal insulation foam. In the simulation model, a constant heat flow can be applied on the bottom surface of the sample and the heat was allowed to pass through the pores as well as the particle skeleton of the Malan loess sample extracted by CT scan section. The side surfaces of the sample are ideal thermal insulators. When the final steady state is reached, the input and output heat fluxes are equal [15].

$$\frac{\varphi_{total}}{S_{in}} = \lambda\frac{T_{in} - T_{out}}{L} \tag{4}$$

Based on the previous expression, the apparent thermal conductivity can be calculated as long as we know the other terms. The experimenter controls the temperatures Tin and Tout and it is easy to determine the total heat flux through the input face by locally using Fourier's law [35,36].

$$\varphi_{total} = \int_{S_{in}} -\lambda_\alpha \vec{\nabla} T_\alpha \cdot \vec{d}s \tag{5}$$

The effective thermal conductivity tensor gives global information about the thermal conduction capabilities of the material. The homogenization theory consists of considering the problem in both a macroscopic domain and a microscopic domain. Here, the macroscopic domain is the periodic domain with characteristic length L and the microscopic domain is the representative elementary volume (REV). The local eigenlength is the period length, and the dimensionless eigenspace variables x* and y* are introduced such that y* = x/l and x* = x/L, where x is the physical space variable [43–45]. As a result of these two eigenvariables, the spatial derivative becomes the following equation:

$$\begin{cases} \nabla_{x*} + \varepsilon^{-1}\nabla_{y*} \\ \varepsilon = \frac{x^*}{y^*} \\ \varepsilon << 1 \end{cases} \tag{6}$$

The unknown T is written as an asymptotic expansion with respect to ε. The Fourier partial differential equation is rewritten considering the new spatial derivatives and the asymptotic expansion of T. Finally, the identification of ε with the same power term leads to the solution of the continuum equation, and the following is the standard equation on the characterization body original.

$$\nabla\left(\lambda_\alpha\left(\vec{\nabla}\vec{b}_\alpha + \vec{\vec{I}}\right)\right) = 0 \tag{7}$$

In Equation (7): $\vec{b}$ is considered as a perturbation of the temperature field and $\vec{b}$ is also verified by Equation (8).

$$\vec{\vec{\lambda}}_{eff} = \frac{1}{V}\sum_{\alpha}\int_{V\alpha}\lambda_\alpha\left(\vec{\nabla}\vec{b}_\alpha + \vec{\vec{I}}\right)dv \tag{8}$$

Malan loess is not homogeneous and contains more than one phase in the process of heat conduction. The properties of each phase can be completely different and the interfaces between the material components play a major role in the global thermal conductivity. Only two conduction media, solid skeletal solid phase and pore structured gas phase, are considered in the dried Malan loess samples, which can be extended to an arbitrary number of phases. Considering the case of a two-phase Malan loess porous media material consisting of a particle skeleton (solid phase) and a pore skeleton (gas phase), defining boundary conditions for each interface between the two phases, the system to be solved is:

$$\begin{cases} \lambda_\xi\nabla^2 T_\xi = 0 \\ T_\alpha = T_\beta \\ -\vec{n}_{\alpha\beta}\cdot\lambda_\alpha\nabla^2 T_\alpha = -\vec{n}_{\alpha\beta}\cdot\lambda_\beta\nabla^2 T_\beta \end{cases} \tag{9}$$

In Equation (9): $\vec{n}_{\alpha\beta}$ is the unit normal vector at the interface from α phase to β phase, and these boundary conditions specify that the normal components of temperature and heat flow are continuous at the interface between the two phases.

3.2.2. Conditions for the Application of the Equation

In order to simulate the heat transfer in porous media in the intact Malan loess 3D model by using the volumetric average theory and heat transfer equation, five assumptions should be set.

(1) The solid skeleton is incompressible ($\rho_s = const$) and is fixed and immobile ($\omega_s = 0$). In this heat conduction simulation, the pore skeleton structure of the intact Malan loess was extracted using CT scan images, and the 3D model parameters of the pore structure could be extracted quantitatively. In order to ensure the expansion of the soil in the Malan loess during the heat conduction experiments, a heat-insulating Styrofoam was fixed between the Malan loess and the PVC pipe, which on the one hand provided thermal insulation and preservation, and on the other hand overcame the expansion of the soil during the heating process and changed the pore skeleton (although this phenomenon was minimal).
(2) The thermal conductivity of the solid and gas phases is a constant, that is $k_i = const(i = s, g)$. Enthalpy: $h(T) = C_pT$, where $C_p = const$. In this study, when

doing the heat transfer experiments of the intact Malan loess, the intact sample was dried and a uniform heat source was added at the bottom to provide a stable heat drive, allowing heat to form thermal convection in the pore skeleton of the Malan loess. And the pore skeleton in the intact Malan loess is extracted unions obtained from CT scans, and only the heat convection process is calculated during the heat transfer simulation, with the thermal conductivity of the gas phase as a constant.

(3) There is no chemical reaction in each phase, and there is no mass source or sink. Ignore the compressibility of gases. The gas phase dissipation is also zero, and $T_l = T_g$. T_l is liquid temperature and T_g is gas temperature.
(4) The surface energy of the solid–gas two-phase interface is negligible.
(5) Near the solid wall $\omega_l = 0, \omega_g = 0$, there is no slip flow occurs.

3.2.3. Solution Method

There are two methods of estimating thermal conductivity when solving for thermal conductivity. The first method is based on experimental simulations of the resolution of the Fourier equation. The second method uses the theory of homogenization over an infinite periodic domain to solve the standard problem derived from the Fourier equation. The first approach is conducted in experimental simulations of thermal conductivity. In addition to the internal interface conditions between the phases, there are three external boundary conditions: (1) A pixel-wide plane of thermal insulator was added to the outer surface of the image along the main direction of heat flux so that the sample could be isolated from the outside world. (2) The input and output (the face perpendicular to the main direction of heat flux) are designed as a pixel-wide plane where the temperature is applied. (3) Any two inputs from the three cases of input temperature, output temperature, and heat flow density are available to be applied to the input and output surfaces, and the simulation selects two inputs depending on the different heat transfer methods, in this case, the input temperature and heat flux are selected this time.

The second approach is to use the theory of homogenization over an infinite periodic domain to solve the standard problem derived from the Fourier equations. This is performed in the thermal conductivity tensor calculation module. The two boundary conditions are generated by the temperature and heat flow along the normal upward component at the interface of the two phases. A periodic boundary condition is applied to the $\vec{b}_\alpha$ geometry to solve the system of Equation (10).

$$\begin{cases} \vec{b_\alpha} = \vec{b_\beta} \\ -\lambda_\alpha \vec{n_{\alpha\beta}} \cdot \left(\vec{\nabla} \vec{b_\alpha} + \vec{\vec{I}} \right) = -\lambda_\beta \vec{n_{\alpha\beta}} \cdot \left(\vec{\nabla} \vec{b_\beta} + \vec{\vec{I}} \right) \end{cases} \quad (10)$$

3.2.4. Discretization of Systems of Equations and System Resolution

The finite volume method is used to solve the system of equations when solving the physical model. The discrete format assumes that the pixels are isotropic. After discretization, the system of equations can be written as $Ax = b$, with A being a sparse symmetric matrix. The system of equations is solved using a fully implicit method (matrix inversion). Iterative solutions are performed using conjugate gradients and ILU preconditioners. The convergence criterion used is the relative reduction of the residual l_2 parametrization.

4. Results and Analysis

4.1. CT Scan Section Processing

At present, in conducting the 3D pore structure characterization and particle quantification analysis platform of porous media, AVIZO software can perform noise reduction processing of CT images, threshold segmentation of images, 3D reconstruction of pores, pore network model construction, pore connectivity analysis, heat conduction simulation, and calculation of absolute thermal conductivity. AVIZO software has a wide range of applications in biomedical, petrology, soil science, and industrial fields and can meet the needs of this study. AVIZO software has a heat conduction module developed based on the volume averaging method, so AVIZO software was chosen to perform image processing, pore network modeling, and heat conduction simulation in this study.

After the calibration of CT sections, the skeleton and pore structure of Malan loess should be extracted, which requires threshold segmentation of CT scan sections of intact Malan loess. The segmentation process is shown in Figure 5. Figure 5a is the original slice. Since the environmental background and internal pores of Malan loess are almost the same in slice imaging, in order to distinguish the internal pores of loess from the external environment, the entire external contour of the Malan loess sample was first segmented by the interactive segmentation method, and the image after segmentation is shown in Figure 5b. The CLOSING module was used to fill the pores inside the entire Malan loess sample so that the pores in Figure 5b can be filled. The slices after filling are shown in Figure 5c, and then a mask layer is generated by the chamfer distance map module as the base for subsequent flood segmentation and top cap segmentation. The mask layer slice is shown in Figure 5d.

Flood segmentation is good for the segmentation of connected macropores in Malan loess samples. Top-cap segmentation can extract the whole internal pores of Malan loess according to the gray value of slice images, but excessive segmentation will occur. The large pores that can be identified by the naked eye were segmented by the flood segmentation method, as shown in Figure 5e. Then, overlay the Figure 5e image and mask layer Figure 5d to obtain a grayscale image, as shown in Figure 5g. The image obtained after the segmentation of the Malan loess sample by top-hat segmentation is shown in Figure 5f. Then, a grayscale image was obtained by superimposing Figure 5f and mask layer Figure 5d, as shown in Figure 5h. The two greyscale images, Figure 5g,h, are superimposed together using the algorithm to obtain the new image, as shown in Figure 5i. Interactive segmentation of the Figure 5i image was performed to obtain the final segmented pore, as shown in Figure 5j. The pores were reconstructed with the segmentation images, and the morphological characteristics of the pores were obtained after statistical analysis, as shown in Figure 5k.

In this study, the porosity of Malan loess was 46% and the specific yield was 9.12%, and the specific yield of intact Malan loess was used as the basis for the division of the threshold value of linked large porosity in the loess. By continuously adjusting the threshold interval of the top hat partition, the final porosity of the model was obtained as 9.116%. The model porosity is smaller than the actual porosity of the Malan loess because pores smaller than 14.089 μm in diameter cannot be identified.

4.2. Pore Network Model

4.2.1. Determination of Representative Volume Units

Considering the limitation of computer computing power and the setting of boundary conditions when doing thermal simulation, and in order to make the space selected for calculation representative, a series of cubes with different side lengths were selected as the representative elementary volume (REV) on the CT scan image in this study, as shown in Figure 6. And the porosity was calculated from the segmented CT scan slice, and when the porosity tends to be stable, the cube was considered as the smallest representative volume unit. The porosity of the representative cube reaches a stable value when the side length of the cube reaches 30 mm, as shown in Figure 7, the subvolume is considered representative of the intact Malan loess sample in this study when the side length of the cube reaches 30 mm or more, and the minimum length of the subvolume unit used for thermal simulation is greater than 30 mm.

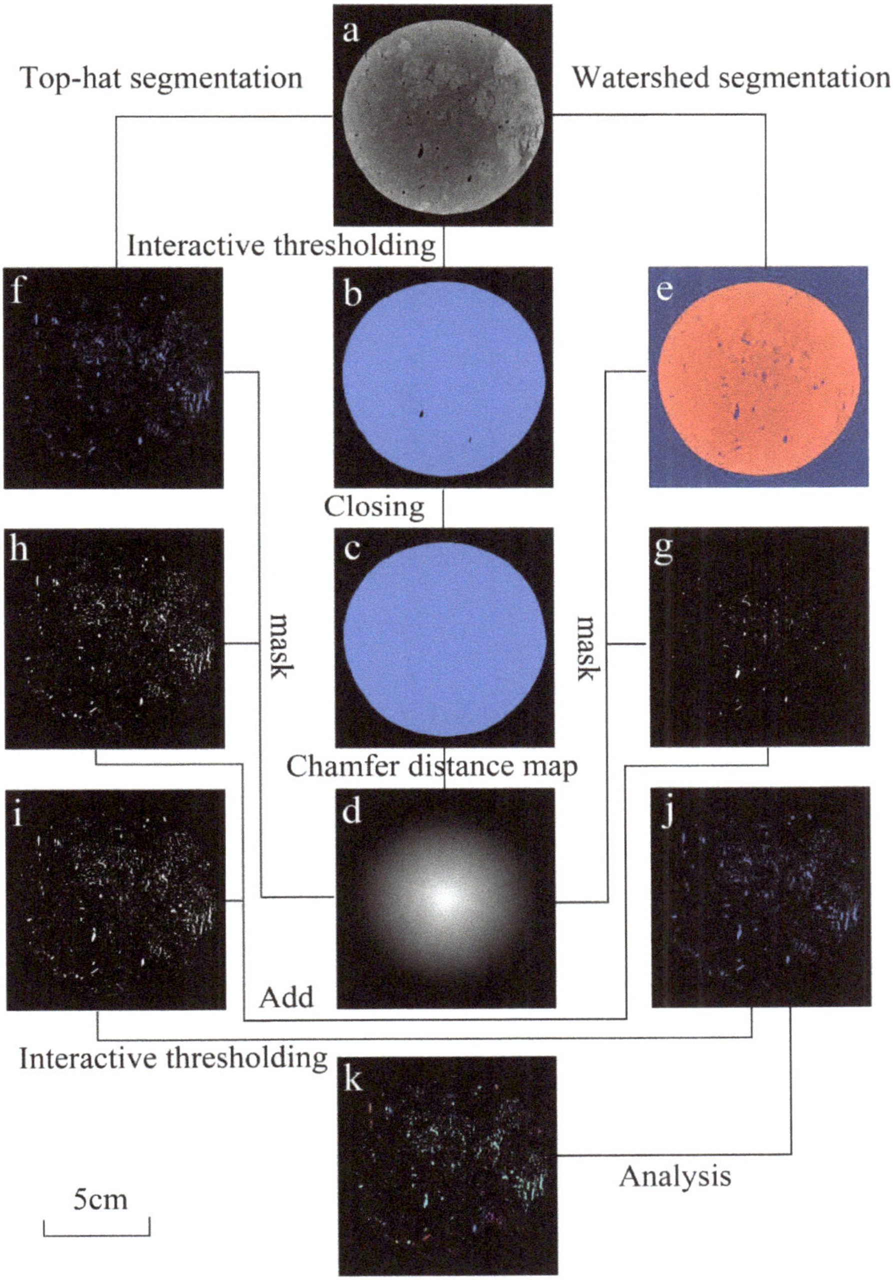

Figure 5. (**a**) Original CT scan section; (**b**) sample contour obtained after interactive segmentation of the original section; (**c**) sample internal pore filling; (**d**) mask layer; (**e**) flood segmentation; (**f**) top-hat segmentation; (**g**) flood segmented pores superimposed with mask layer; (**h**) top-hat segmented pores superimposed with mask layer; (**i**) g image and h image overlay; (**j**) pore space after splitting; (**k**) pore analysis.

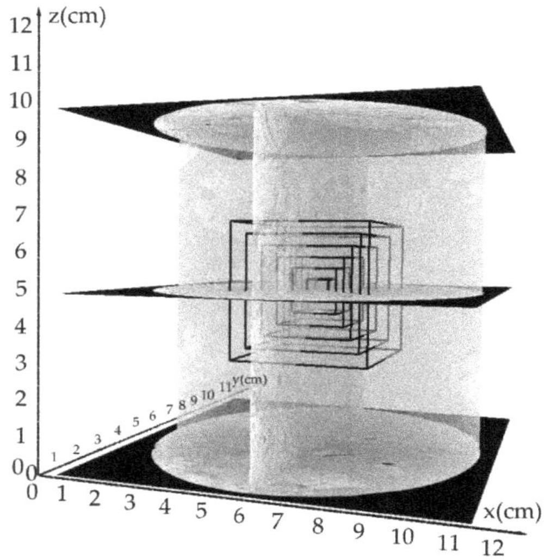

Figure 6. Schematic diagram of the selection of the volume of the characterization unit.

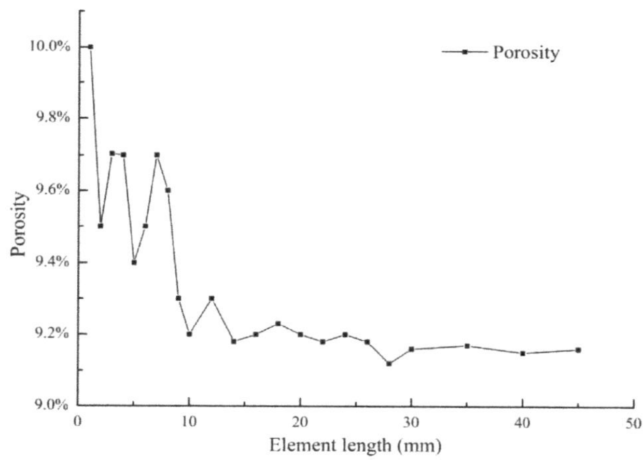

Figure 7. Porosity versus cube side length.

4.2.2. Construction of Pore Network Model

Because the morphological structure of the interconnected pores in the intact loess is very complex, the pore network model of the interconnected pores in the intact Malan loess can be established by CT scan sections after segmentation. The pore network model assumes that an interconnected pore has multiple pore network models consisting of different pores connected by pore channels, and the pore throat was the location with the smallest equivalent diameter in the pore channel. The established pore network model can be represented by a ball-and-stick model, where the stick represents the pore channel, and the ball represents the pore throat. The pore network model can be used to quantify the pores in the intact Malan loess. Key information such as the length of the pore channel, the area of the pore howl, and the number of allotments are obtained. The pore network model is shown in Figure 8.

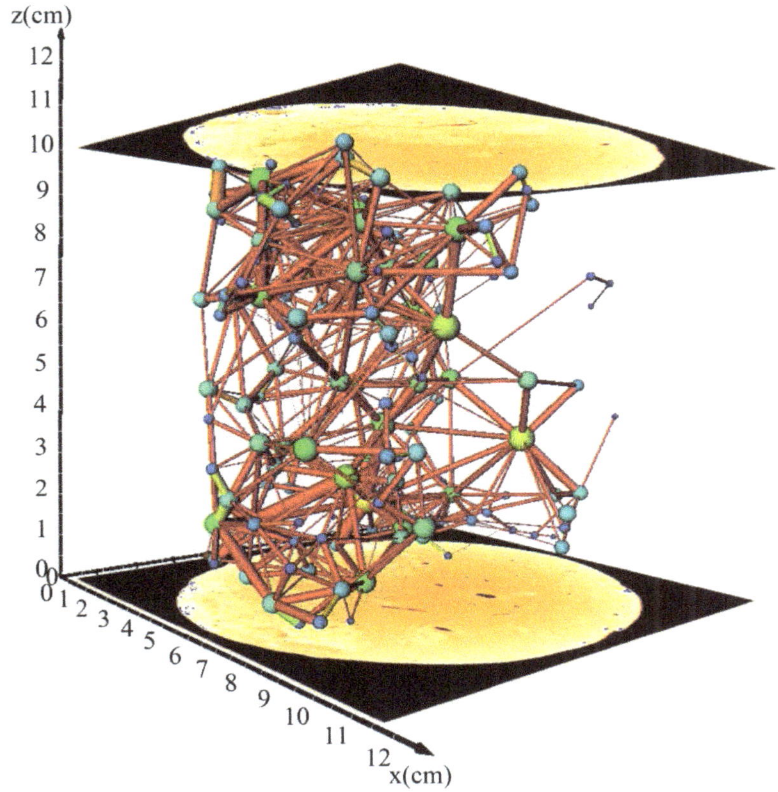

Figure 8. Pore network model.

4.3. Pore Network Model Analysis

4.3.1. Analysis of the Distribution Characteristics of Pores and Pore Throats

According to the established pore network model, the distribution range of the pore throat and pore channel equivalent diameter was counted, and the distribution range of the pore was wider than that of the pore throat. Histogram statistics of the equivalent diameter of pore and pore throat with 200 μm interval were conducted, and it was found that the distribution percentage of pore and pore roar approximately obeyed the lognormal distribution. According to Figure 9, the pores were mainly distributed in the interval of pore diameter of 800–6600 μm, accounting for 90% of the total. The pore throats were mainly distributed in the interval of pore diameter of 200–1800 μm, accounting for 96.62% of the total.

4.3.2. Pore Connectivity Characteristics Analysis

The established pore network model can be used to calculate the coordination of the number of pore throats connected to the pores in the ball-and-stick model. The coordination number can characterize the connectivity of the pores, and a higher coordination number represents better connectivity with the surrounding pores. According to the constructed pore network model, the calculated coordination number was statistically analyzed, as shown in Figure 10. The maximum number of coordination is 32, the plurality of coordination was 3 and 6, and the coordination number was mainly distributed between 1 and 12, which can indicate that the overall connectivity of pores was good.

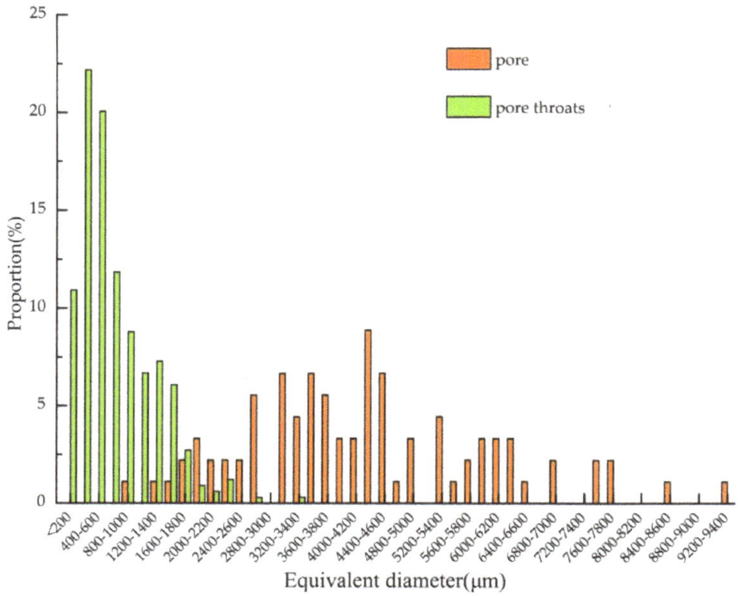

Figure 9. Pore and pore throat distribution characteristics.

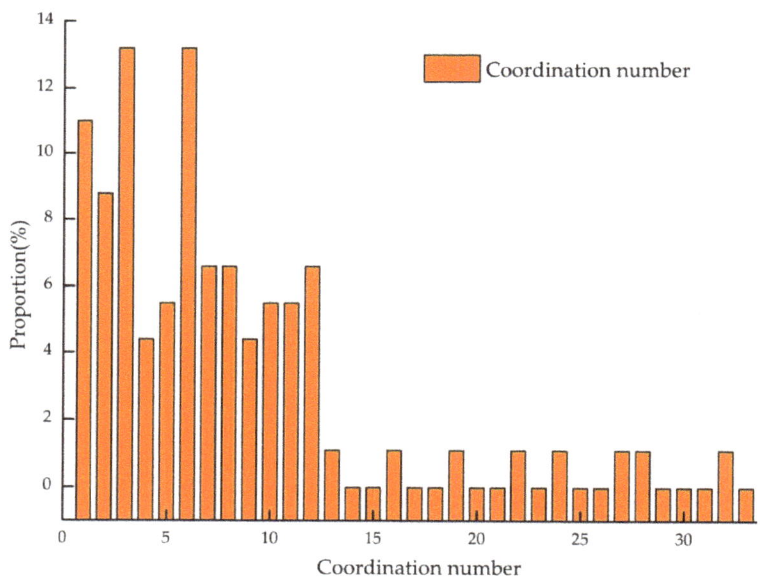

Figure 10. Distribution of pore coordination number.

4.4. Analysis of Heat Conduction Experimental Results

4.4.1. Analysis of Heat Conduction Experimental Verification Results

The validation test continued to follow the original heating process on the bottom surface of the Malan loess sample after perforation, and the temperature control knob was adjusted to 100 °C. The temperature change on the surface of the Malan sample was observed by the thermal imager. If the heat conduction in Malan loess is carried out along the pores first, then eight regular abnormal temperature points can be observed at the location of the perforated Malan loess surface, as shown in Figure 11. There was no abnormal temperature point on the surface of the sample at the beginning of the observation, and the temperature of the sample surface is 16.6 °C as the surrounding temperature. The temperature of the abnormal temperature point is 21.3 °C, which is higher than the temperature of other points on the surface of the sample, and the temperature of this abnormal temperature point increases until the end of the heating process. At 49 min of heating, an abnormal temperature point was observed in the borehole with a diameter of 1.5 mm, and at 59 min of heating, an abnormal temperature point was observed in the borehole with a diameter of 1 mm. During the heating process up to 78 min, an ectothermic point can be observed in the borehole with a 0.5 mm diameter. During the heating process, there were also ectothermic spots at the pore exits on the surface of the Malan loess sample itself, and the temperatures of these eight boreholes were higher than those of other spots on the sample surface during the subsequent heating process. Finally, after 300 min, the temperature on the surface of the Malan loess sample hardly changed and the whole heat transfer process stabilized.

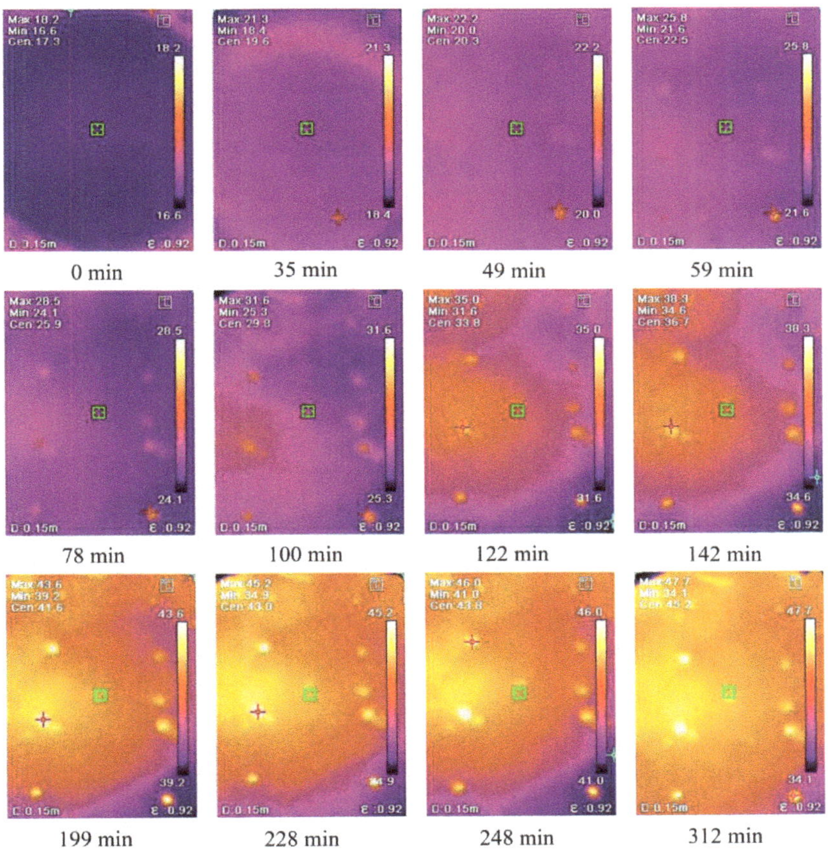

Figure 11. Thermal imaging of the surface of the perforation experiment.

This experiment proves that when an external temperature pressure is applied to the Malan loess, the heat transfer in the solid and gas phase media of Malan loess generally decreases in the direction of heat flow, and the heat is transferred in the pores inside the Malan loess first, and then the heat brought by the pores heats up the pore walls, thus increasing the temperature of the solid skeleton and the temperature of the whole sample. It can be found that the larger the borehole diameter is, the earlier the appearance of the abnormal temperature point is, so it can be verified that the heat conduction in the Malan loess sample is conducted along the pore space first and then the heat is transferred from the heated pore wall to the surrounding soil skeleton.

4.4.2. Analysis of Heat Conduction Experimental Results

The thermal imaging observation device of Malan loess was connected, and the surface temperature surface of the Malan loess sample was observed, and the observation results are shown in Figure 12. After measuring, the ambient temperature is 18 °C. The initial imaging process is shown as 0 min in Figure 12, the temperature of the sample surface is about 19 °C, there is no abnormal temperature point on the surface of the Malan loess sample in the initial stage, and the first abnormal temperature point is found after 45 min. Finally, the temperature of the surface of the Maran loess sample hardly changed after 128 min, and no new abnormal temperature points appeared, and the whole heat transfer process was stabilized.

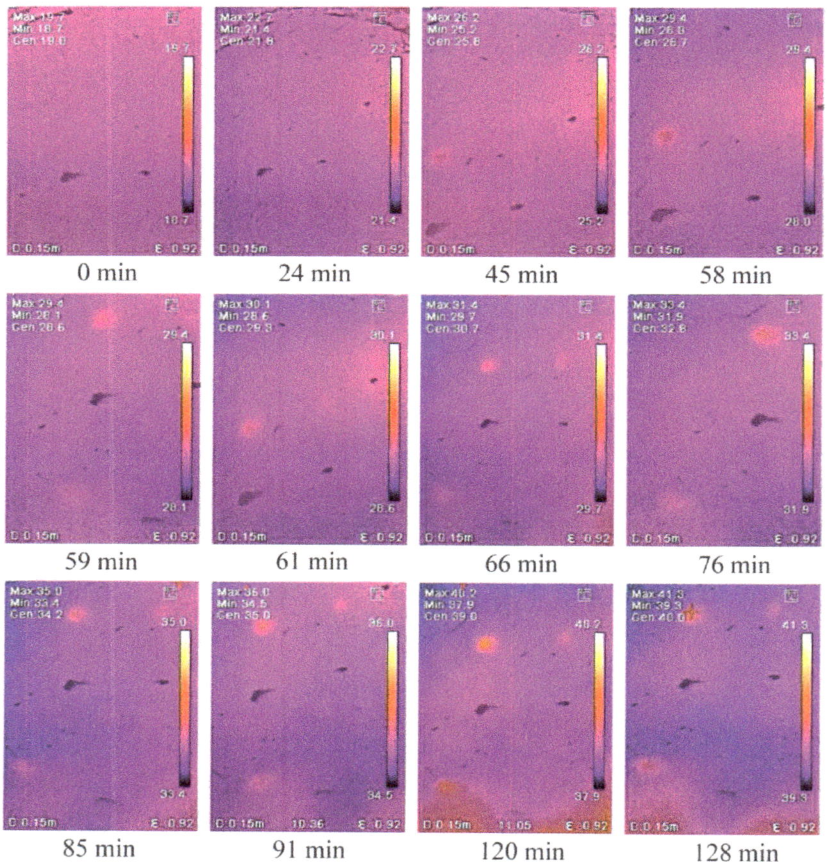

Figure 12. Thermal imaging of Malan loess surface.

4.4.3. Heat Conduction Experiment Abnormal Temperature Point Extraction

The abnormal temperature points on the surface of the Malan loess sample were extracted from the thermogram, and all the observed abnormal temperature points were projected onto the image of the intact Malan loess surface as shown in Figure 13a, and then these points were projected onto the CT scan section of the Malan loess sample surface for repositioning as shown in Figure 13b. Figure 13a shows the point distribution of the abnormal temperature points on the surface of the Malan loess sample, and Figure 13b shows the projection of the abnormal temperature points on the CT scan image. The 3D pore structure of these abnormal temperature points can be searched by the projected points on the CT scan section and the reconstructed 3D pore model. The extracted three-dimensional pore model is shown in Figure 13c. Figure 13c shows the plan view of the corresponding pore positions of these five abnormal temperature points, and Figure 13d shows the three-dimensional shapes of the five abnormal temperature points. The longest pore was 9.13 cm, the shortest pore was 2.86 cm, and the average length is 5.53 cm. The maximum pore inclination was 84°, the minimum inclination is 52°, and the average inclination was 76°. The actual length of the pore and the coordinates of the beginning and ending points of the pore were extracted from the three-dimensional pore model. The tortuosity of the pore was obtained by dividing the actual pore length by the linear distance between the two ends of the pore. The maximum tortuosity of the pore was 1.0051, the minimum tortuosity was 0.9692, the average tortuosity was 0.9886, the pores were in the form of a dendritic, and the pore was more strongly connected.

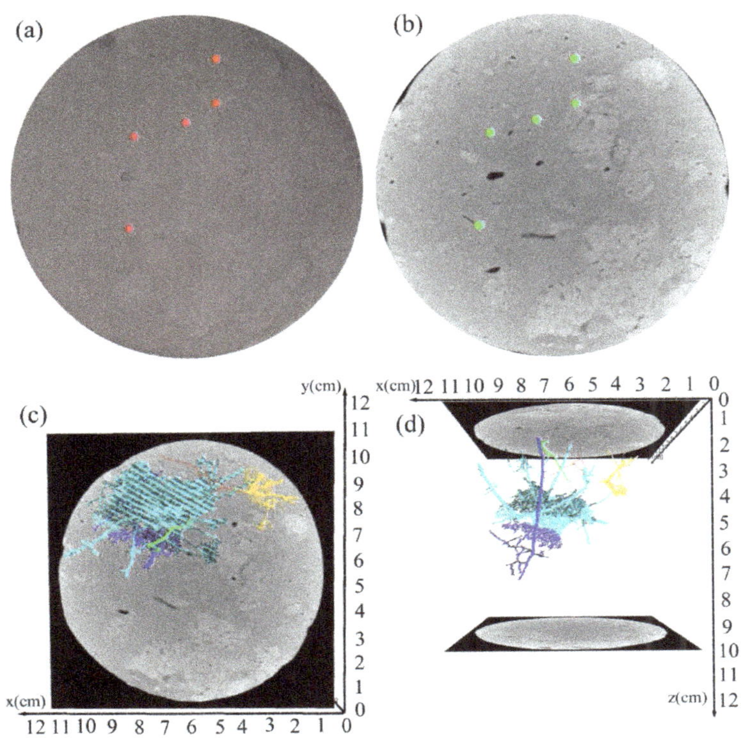

Figure 13. Sample surface abnormal temperature point localization. (**a**) The abnormal temperature points were projected onto the image of the intact Malan loess surface; (**b**) The abnormal temperature points on the CT scan image; (**c**) The corresponding pore positions of these five abnormal temperature points; (**d**) The three-dimensional shapes of the five abnormal temperature points.

4.5. Analysis of Simulation Results

The intact Malan loess sample was a cylinder with a height of 10 cm and a diameter of 10 cm. Due to the influence of computer operation, the simulation was carried out by using the method of characterizing representative elementary volume (REV). Select a cube that can cover the abnormal temperature points, and the height of the selected cube was 3 cm. In the heat transfer experiment, the bottom input temperature was 80 °C, and the surface output temperature was stable when it was 40 °C. Assuming that the overall temperature transfer mode is linearly decreasing, the heat transfer will decrease by 4 °C every 1 cm, the heat transfer will decrease by 3 cm, and the temperature will decrease by 12 °C, as shown in Figure 14a. Therefore, the input temperature of the bottom surface of the simulation cube was set to 52 °C during the simulation. According to the ratio of the input power to the input area, the heat flux was set to 0.0002 $W{\cdot}m^{-2}$, and the surrounding was set to the heat insulation boundary. The thermal conductivity of the air was set to 2.59×10^{-2} $W{\cdot}m^{-1}{\cdot}K^{-1}$, and the thermal conductivity of the skeleton of the intact Malan loess sample was set to 0.32 $W{\cdot}m^{-1}{\cdot}K^{-1}$. The simulation results are shown in Figure 14b, where the heat follows the conduction direction and the temperature decreases continuously, and the temperature varies with the pore morphology at the same level, and the heat is preferentially conducted along the pores, and then the heat is transferred to the solid skeleton particles. Among them, Figure 14c shows the 41 °C isotherm in the heat conduction process, where Figure 14d shows the 45 °C isotherm in the heat conduction process, and where Figure 14e shows the 50 °C isotherm in the heat conduction process. It can be found that the more complex the pore morphology is, the more abnormal temperature points are formed in the vertical plane along the direction of heat flow in the process of heat conduction. The temperature gradient changes less with the growth of the path in the process of vertical heat transfer in Malan loess.

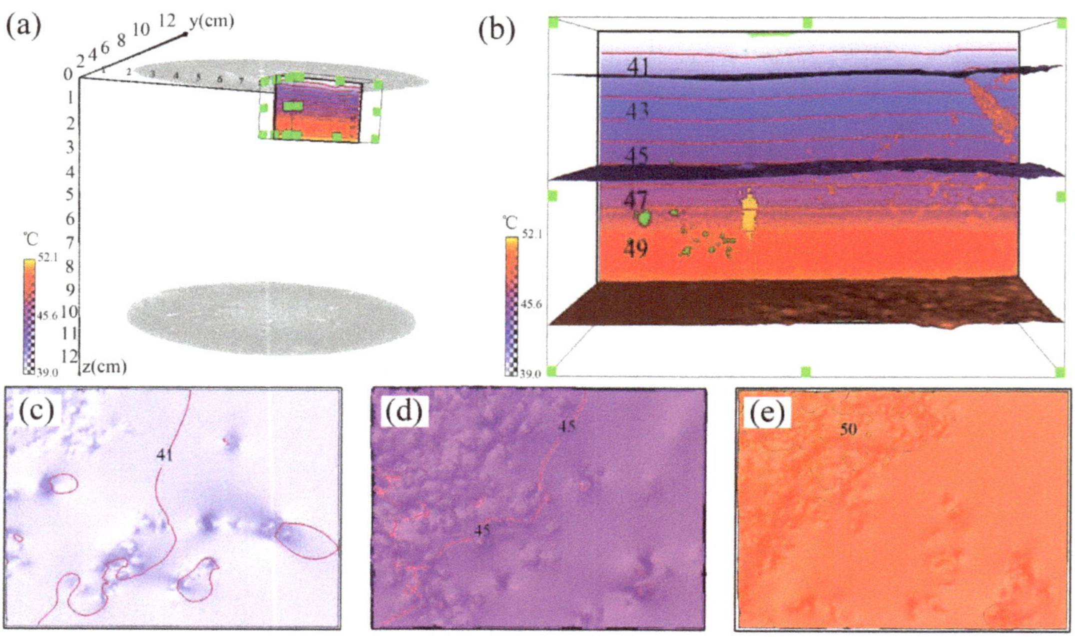

Figure 14. Simulation results of heat transfer simulation. (**a**) Three-dimensional simulation model; (**b**) The simulation results; (**c**) The 41 °C isotherm in the heat conduction process; (**d**) The 45 °C isotherm in the heat conduction process; (**e**) The 50 °C isotherm in the heat conduction process.

The calculated isotherm plot of 41.2 °C on the surface of the Malan loess sample after the steady state is shown in Figure 15. The red circles in Figure 15 show the distribution of abnormal temperature points on the simulated surface of the heat conduction of the intact Malan loess, which is highly consistent with the actual observed abnormal temperature points in Figure 13, which fully proves that the heat conduction of the native Malan loess starts preferentially along the pores in the process of heat conduction and then is transferred to the solid skeletal particles.

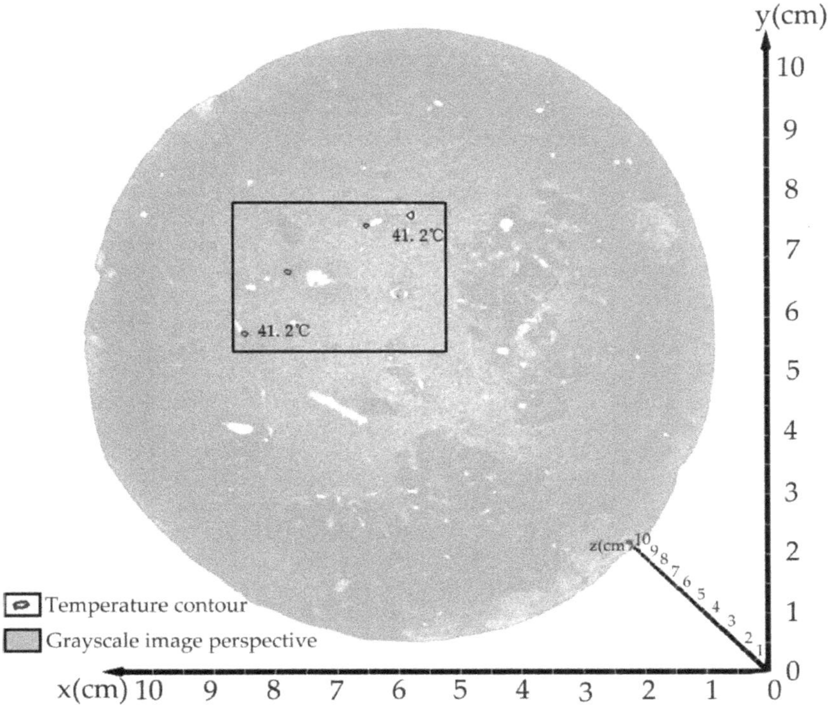

Figure 15. Heat conduction simulation results in surface abnormal temperature points.

5. Discussion

In order to quantitatively characterize the macropore structure inside the intact Malan loess, a cylindrical intact Malan loess sample with a diameter of 10 cm and a height of 10 cm was scanned using industrial computed tomography, and the CT scan section was segmented using a composite segmentation method. Usually, the segmentation methods such as interactive segmentation, flood segmentation, and top-hat segmentation can segment the pores and the particle skeleton in the samples, but the principles of different segmentation methods are different, and over-segmentation or under-segmentation can occur. Therefore, the overlay of different segmentation methods should be considered to eliminate this situation due to over-segmentation or under-segmentation. In this study, we use the characteristics of flood segmentation to segment large porosity and top-hat segmentation to segment small porosity. By overlaying the mask layer, we use the large porosity segmented by flood segmentation as the base and superimpose the small porosity segmented by top-hat segmentation as the final segmented image, which can better solve the image segmentation problem. The porosity of the segmented sample is 9.116%, and the specific yield is treated as the large porosity, and this segmentation method satisfies the large porosity characteristics found in the heat transfer experiment in this study [1].

Malan loess is a porous medium with strong verticality and spatial anisotropy. In order to study the heat transfer process of Malan loess porous medium, the morphological parameters of the pore skeleton in Malan loess samples need to be extracted, and the morphological parameters of Malan loess samples were extracted by CT grayscale slices of Malan loess samples through computed tomography. The pore morphology of Malan loess samples was found to be complex, with strong pore development in the vertical direction and strong overall connectivity. In this study, only the large pores with pore sizes larger than 200 μm were focused on. A 140.349 μm resolution can meet the study requirements. In the future, more in-depth characterization of the loess structure requires more scales and higher resolution CT scans to accurately characterize the pore structure within the loess [1,2,5].

In order to study the heat transfer process of Malan loess, a non-contact thermal imager was used to observe the surface of the Malan loess sample. This observation method does not damage the pore structure of the soil and can detect the temperature of all points on the surface of the sample at the same time. However, it was found that the change in temperature on the surface of the intact Malan loess samples was not homogeneous, and a series of abnormal temperature points appeared. This phenomenon is consistent with the results of the simulation process of the particle-to-fluid heat transfer process studied by Urrutia et al. [20]. This study aimed to investigate whether the abnormal temperature points detected by the surface thermal imager of Malan loess samples are caused by the heterogeneity of the particle skeleton or by the conductivity of different media in multiphase porous media. In this study, a perforation verification experiment was designed to artificially create a series of regular pores in the Malan loess sample, and observed by a thermal imager, it was found that abnormal temperature points appeared at the contact point between the artificial regular pores and the surface of the Malan loess sample, and the temperature at the pores was higher than the temperature at the solid skeleton by about 1 °C. This verifies from the side that in the heat transfer process of Malan loess, heat is preferentially conducted along the large pores at a given temperature and pressure, and then the heat is transferred to the surrounding solid skeleton particles. This experimental method can be applied to other porous media samples, and the process of heat conduction in different porous media can be studied [12,15,16].

In the study of the heat transfer process in porous media, the conceptual model of porous media is usually generalized, but the morphological parameters of porous media in nature are very complex. In this study, CT scan sections of Malan loess samples were obtained by computerized tomography, and the pore skeleton of the Malan loess sample was reconstructed to extract the pore network model parameters of the linked pores in Malan loess samples, which provided a conceptual model for heat transfer simulation. Through heat conduction simulation, temperature anomalies appear on the surface of the selected area of the simulation, and the results of the heat conduction simulation are found to be in high agreement with the actual experimental observations. The heat conduction simulation allows the calculation of heat transfer processes in porous media with very small scales, and the soil temperature variation parameters can be calculated in the form of computer simulations for soils where soil moisture temperature sensors are not well placed [18,19].

6. Conclusions

The following conclusions can be drawn from the heat conduction simulation based on the reconstruction of the internal pores of intact Malan loess and the comparative analysis of the experimental results of heat conduction thermography of intact Malan loess.

The internal image sections of intact Malan loess samples were obtained by CT scanning, and a model of the fine-scale pore structure of Malan loess was reconstructed using these CT scanned sections. A device was designed to observe the fine-scale heat transfer in Malan loess and the thermal insulation conditions were set on the side surfaces of the cylindrical Malan loess samples. The temperature pressure was applied at the bottom of the sample, and the change in the surface temperature of the Malan loess sample was observed at the top of the sample with a thermal imager. The three-dimensional pore structure extracted by the volume averaging method combined with a CT scan was used to simulate the process of heat transfer in the solid and gas phase media of the Malan loess sample. The experimental results were compared and analyzed with the simulation results to verify the reliability of the simulation results.

By setting up the verification experiment of perforation of Malan loess heat transfer in the solid and gas medium preferentially, it can be seen through the perforation experiment that in the process of heat transfer of intact Malan loess, the heat was preferentially transferred along the large pores, and an abnormal temperature point was formed at the junction of the surface and internal pores of Malan loess, and the temperature of the jointed large pores at the surface of the sample was about 1 °C higher than that of the surrounding solid skeleton.

The simulation process of heat transfer can be performed based on the volume averaging method combined with the pore skeleton extracted by CT scanning, and the thermal conductivity can also be calculated, providing a new solution for the calculation of thermal conductivity, and the heat transfer process of small-scale Malan loess can be studied to provide parameters such as thermal conductivity for heat transfer processes that are difficult to observe at the macroscopic level.

In this thermal imaging experiment of intact Malan loess, the smallest pore diameter detected was 500 μm and the minimum error in temperature was ±0.2 °C minimum. In order to detect smaller pores, a thermal imager with higher accuracy and a CT scanning device with higher resolution can be used. If smaller-scale pore loess heat transfer studies are conducted, smaller soil samples and computers with higher computing power are required. With the future development of computer technology, the improvement of CT scanning technology and infrared thermal imaging technology, smaller sample size, smaller scale, and higher resolution soil heat transfer studies can be carried out, so that the influence of loess pore structure and loess particle skeleton on the heat transfer of loess porous media at different scales can be better studied. With the development of a simulation model of heat transfer in Malan loess at multiple scales, coupling the heat conduction properties of macro-scale Malan loess with the microstructure of Malan loess, the micro-mechanism of the heat conduction process in Malan Loess can be better understood. Future research will further explore the relationship between pore structure and heat conduction and will provide scientific basis and technical support for the optimization and application of heat transfer properties of Malan Loess as well as other porous media.

Author Contributions: Conceptualization, Y.L. (Yangchun Lu) and Y.L. (Yudong Lu); methodology, Y.L. (Yangchun Lu); software, Y.L. (Yangchun Lu); validation, Y.L. (Yangchun Lu) and Y.L. (Yudong Lu); formal analysis, Y.L. (Yangchun Lu) and T.L.; investigation, Y.L. (Yangchun Lu), T.L. and X.Z.; resources, G.Z. and B.W.; data curation, Y.L. (Yangchun Lu) and X.Z.; writing—original draft preparation, Y.L. (Yangchun Lu); writing—review and editing, Y.L. (Yangchun Lu); visualization, Y.L. (Yangchun Lu); supervision, Y.L. (Yudong Lu). All authors have read and agreed to the published version of the manuscript.

Funding: This research was funded by the National Natural Science Foundation of China (Grants No. U2243204, 41630634, 42261144749 and 41877232), and the Natural Science Foundation of Hebei Province (Grant No.: E2021210092).

Data Availability Statement: Not applicable.

Acknowledgments: The editor and reviewers are highly appreciated for providing positive and constructive comments and suggestions concerning this manuscript.

Conflicts of Interest: The authors declare no conflict of interest. The funders had no role in the design of the study; in the collection, analyses, or interpretation of data; in the writing of the manuscript; or in the decision to publish the results.

Nomenclature

$\vec{\varphi}$	Heat flow ($W \cdot m^{-2}$)
λ	Thermal conductivity of the material ($W \cdot m^{-1} \cdot K^{-1}$)
T	Temperature (K)
$(\rho c_p)_\alpha$	Heat capacity of the α phase ($J \cdot m^{-3} \cdot K^{-1}$)
ρ_α	Density of the α phase ($kg \cdot m^{-3}$)
$(c_p)_\alpha$	Specific heat capacity of the α phase ($J \cdot kg^{-1} \cdot K^{-1}$)
λ_α	Thermal conductivity of the α phase ($W \cdot m^{-1} \cdot K^{-1}$)
T_α	Temperature of the α phase (K)
φ_{total}	Total heat flux through the input surface ($W \cdot m^{-2}$)
S_{in}	Area of the material in contact with the heat input surface (m^2)
λ	Apparent thermal conductivity of the material ($W \cdot m^{-1} \cdot K^{-1}$)
T_{in}	Input temperatures (K)
T_{out}	Output temperatures (K)
L	Length of the material sample (m)
λ_α	Thermal conductivity in any phase (α phase)
T_α	Temperature in any phase (α phase)
S_{in}	Area of the material in contact with the heat input surface (m^2)
$\vec{b}$	Considered as a perturbation of the temperature field
$\overrightarrow{\overrightarrow{\lambda}}_{eff}$	Effective thermal conductivity tensor
V	Total volume of the sample (m^3)
α	A conduction phase
V_α	Volume of each conduction phase (m^3)
$\vec{n}_{\alpha\beta}$	Unit normal vector at the interface from α phase to β phase

References

1. Li, X.; Lu, Y.; Zhang, X.; Fan, W.; Lu, Y.; Pan, W. Quantification of macropores of Malan loess and the hydraulic significance on slope stability by X-ray computed tomography. *Environ. Earth Sci.* **2019**, *78*, 522. [CrossRef]
2. Li, Y.; Mo, P.; Wang, Y.; Zhang, T.; Zhang, H. Strength anisotropy of Malan loess and the implications for the formation of loess walls and columns. *Catena* **2020**, *194*, 104809. [CrossRef]
3. Lu, Q.; Qiao, J.; Peng, J.; Liu, Z.; Liu, C.; Tian, L.; Zhao, J. A typical Earth fissure resulting from loess collapse on the loess plateau in the Weihe Basin, China. *Eng. Geol.* **2019**, *259*, 105189. [CrossRef]
4. Yuan, K.; Ni, W.; Lü, X.; Zhu, M.; Wang, H.; Nie, Y. Mechanical properties and microstructure evolution of Malan loess. *Quat. Int.* **2022**, *637*, 74–84. [CrossRef]
5. Zhang, X.; Lu, Y.; Li, X.; Lu, Y.; Sun, J.; Pan, W. Multilevel Collapsibility of Loess under Irrigation in Jinya Town, Gansu Province, China. *Adv. Civ. Eng.* **2019**, *2019*, 2153679. [CrossRef]
6. Lu, Y.; Lu, Y.; Lu, T.; Wang, B.; Zeng, G.; Zhang, X. Computing of Permeability Tensor and Seepage Flow Model of Intact Malan Loess by X-ray Computed Tomography. *Water* **2023**, *15*, 2851. [CrossRef]
7. Li, Y.; Zhang, T.; Zhang, Y.; Xu, Q. Geometrical appearance and spatial arrangement of structural blocks of the Malan loess in NW China: Implications for the formation of loess columns. *J. Asian Earth Sci.* **2018**, *158*, 18–28. [CrossRef]
8. Luo, H.; Wu, F.; Chang, J.; Xu, J. Microstructural constraints on geotechnical properties of Malan Loess: A case study from Zhaojiaan landslide in Shaanxi province, China. *Eng. Geol.* **2018**, *236*, 60–69. [CrossRef]
9. Zhang, X.; Li, X.; Lu, Y.; Lu, Y.; Fan, W. A study on the collapse characteristics of loess based on energy spectrum superposition method. *Heliyon* **2023**, *9*. [CrossRef]
10. Chen, Y.; Shen, C.; Lu, P.; Huang, Y. Role of pore structure on liquid flow behaviors in porous media characterized by fractal geometry. *Chem. Eng. Process. Process Intensif.* **2014**, *87*, 75–80. [CrossRef]
11. Pia, G.; Sanna, U. Case studies on the influence of microstructure voids on thermal conductivity in fractal porous media. *Case Stud. Therm. Eng.* **2014**, *2*, 8–13. [CrossRef]
12. Qin, X.; Cai, J.; Xu, P.; Dai, S.; Gan, Q. A fractal model of effective thermal conductivity for porous media with various liquid saturation. *Int. J. Heat Mass Transf.* **2019**, *128*, 1149–1156. [CrossRef]
13. Zheng, C.; Šimůnek, J.; Zhao, Y.; Lu, Y.; Liu, X.; Shi, C.; Li, H.; Yu, L.; Zeng, Y.; Su, Z. Development of the Hydrus-1D freezing module and its application in simulating the coupled movement of water, vapor, and heat. *J. Hydrol.* **2021**, *598*, 126250. [CrossRef]
14. To, Q.D.; Bonnet, G. Fourier transform approach to homogenization of frequency-dependent heat transfer in porous media. *Int. J. Numer. Methods Heat Fluid Flow* **2023**, *33*, 2023–2048. [CrossRef]
15. Wang, J.; Carson, J.K.; North, M.F.; Cleland, D.J. A new approach to modelling the effective thermal conductivity of heterogeneous materials. *Int. J. Heat Mass Transf.* **2006**, *49*, 3075–3083. [CrossRef]

16. Wang, Y.; Ma, C.; Liu, Y.; Wang, D.; Liu, J. A model for the effective thermal conductivity of moist porous building materials based on fractal theory. *Int. J. Heat Mass Transf.* **2018**, *125*, 387–399. [CrossRef]
17. Yu, B.; Cheng, P. Fractal Models for the Effective Thermal Conductivity of Bidispersed Porous Media. *J. Thermophys. Heat Transf.* **2002**, *16*, 22–29. [CrossRef]
18. Cheng, Q.; Tang, C.-S.; Lin, Z.-Z.; Tian, B.-G.; Shi, B. Measurement of water content at bare soil surface with infrared thermal imaging technology. *J. Hydrol.* **2022**, *615*, 128715. [CrossRef]
19. Zhao, W.; Dong, X.; Wu, Z.; Wei, C.; Li, L.; Yu, D.; Fan, X.; Ma, Y. Using infrared thermal imaging technology to estimate the transpiration rate of citrus trees and evaluate plant water status. *J. Hydrol.* **2022**, *615*, 128671. [CrossRef]
20. Reyes-Urrutia, A.; Venier, C.; Mariani, N.J.; Nigro, N.; Rodriguez, R.; Mazza, G. A CFD Comparative Study of Bubbling Fluidized Bed Behavior with Thermal Effects Using the Open-Source Platforms MFiX and OpenFOAM. *Fluids* **2021**, *7*, 1. [CrossRef]
21. Kou, J.; Liu, Y.; Wu, F.; Fan, J.; Lu, H.; Xu, Y. Fractal analysis of effective thermal conductivity for three-phase (unsaturated) porous media. *J. Appl. Phys.* **2009**, *106*, 054905. [CrossRef]
22. Nait-Ali, B.; Haberko, K.; Vesteghem, H.; Absi, J.; Smith, D.S. Thermal conductivity of highly porous zirconia. *J. Eur. Ceram. Soc.* **2006**, *26*, 3567–3574. [CrossRef]
23. Raed, K.; Gross, U. Modeling of Influence of Gas Atmosphere and Pore-Size Distribution on the Effective Thermal Conductivity of Knudsen and Non-Knudsen Porous Materials. *Int. J. Thermophys.* **2009**, *30*, 1343–1356. [CrossRef]
24. Chen, Y.-C. Non-Darcy flow stability of mixed convection in a vertical channel filled with a porous medium. *Int. J. Heat Mass Transf.* **2004**, *47*, 1257–1266. [CrossRef]
25. Babaei, H.; McGaughey, A.J.H.; Wilmer, C.E. Effect of pore size and shape on the thermal conductivity of metal-organic frameworks. *Chem. Sci.* **2017**, *8*, 583–589. [CrossRef]
26. Brzić, D.; Pešić, R.; Arsenijević, Z.; Đuriš, M.; Bošković-Vragolović, N.; Kaluđerović-Radoičić, T. Heat transfer to a sphere immersed in a fluidized bed of coarse particles with transition from bubbling to turbulent flow regime. *Part. Sci. Technol.* **2022**, *41*, 75–83. [CrossRef]
27. Freitas, S.K.S.; Borges, R.S.; Merlini, C.; Barra, G.M.O.; Esteves, P.M. Thermal Conductivity of Covalent Organic Frameworks as a Function of Their Pore Size. *J. Phys. Chem. C* **2017**, *121*, 27247–27252. [CrossRef]
28. Jin, H.; Zhang, M.; Shen, S. Heat Transfer around and through Multiple Porous Particles. *Ind. Eng. Chem. Res.* **2022**, *61*, 8269–8278. [CrossRef]
29. Carmignato, S.; Dewulf, W.; Leach, R. Industrial X-Ray Computed Tomography | | Applications of CT for Non-destructive Testing and Materials Characterization. *Ind. X-ray Comput. Tomogr.* **2018**, *17*, 267–331.
30. Ivanov, A.L.; Shein, E.V.; Skvortsova, E.B. Tomography of Soil Pores: From Morphological Characteristics to Structural–Functional Assessment of Pore Space. *Eurasian Soil Sci.* **2019**, *52*, 50–57. [CrossRef]
31. Samant, P.; Trevisi, L.; Ji, X.; Xiang, L. X-ray induced acoustic computed tomography. *Photoacoustics* **2020**, *19*, 100177. [CrossRef]
32. Shanti, N.O.; Chan, V.W.L.; Stock, S.R.; De Carlo, F.; Thornton, K.; Faber, K.T. X-ray micro-computed tomography and tortuosity calculations of percolating pore networks. *Acta Mater.* **2014**, *71*, 126–135. [CrossRef]
33. Zhao, Z.; Zhou, X.-P. An integrated method for 3D reconstruction model of porous geomaterials through 2D CT images. *Comput. Geosci.* **2019**, *123*, 83–94. [CrossRef]
34. Weigand, T.M.; Schultz, P.B.; Giffen, D.H.; Farthing, M.W.; Crockett, A.; Kelley, C.T.; Gray, W.G.; Miller, C.T. Modeling Nondilute Species Transport Using the Thermodynamically Constrained Averaging Theory. *Water Resour. Res.* **2018**, *54*, 6656–6682. [CrossRef]
35. Wood, B.D.; Valdés-Parada, F.J. Volume averaging: Local and nonlocal closures using a Green's function approach. *Adv. Water Resour.* **2013**, *51*, 139–167. [CrossRef]
36. Yang, C.; Huang, R.; Lin, Y.; Qiu, T. Volume averaging theory (VAT) based modeling for longitudinal mass dispersion in structured porous medium with porous particles. *Chem. Eng. Res. Des.* **2020**, *153*, 582–591. [CrossRef]
37. Battiato, I.; Ferrero, V.P.T.; O' Malley, D.; Miller, C.T.; Takhar, P.S.; Valdés-Parada, F.J.; Wood, B.D. Theory and Applications of Macroscale Models in Porous Media. *Transp. Porous Media* **2019**, *130*, 5–76. [CrossRef]
38. Blunt, M.J.; Bijeljic, B.; Dong, H.; Gharbi, O.; Iglauer, S.; Mostaghimi, P.; Paluszny, A.; Pentland, C. Pore-scale imaging and modelling. *Adv. Water Resour.* **2013**, *51*, 197–216. [CrossRef]
39. Bu, S.S.; Yang, J.; Zhou, M.; Li, S.Y.; Wang, Q.W.; Guo, Z.X. On contact point modifications for forced convective heat transfer analysis in a structured packed bed of spheres. *Nucl. Eng. Des.* **2014**, *270*, 21–33. [CrossRef]
40. Huo, A.; Wang, X.; Zhao, Z.; Yang, L.; Zhong, F.; Zheng, C.; Gao, N. Risk Assessment of Heavy Metal Pollution in Farmland Soils at the Northern Foot of the Qinling Mountains, China. *Int. J. Environ. Res. Public Health* **2022**, *19*, 14962. [CrossRef]
41. Huo, A.; Yang, L.; Luo, P.; Cheng, Y.; Peng, J.; Nover, D. Influence of landfill and land use scenario on runoff, evapotranspiration, and sediment yield over the Chinese Loess Plateau. *Ecol. Indic.* **2021**, *121*, 107208. [CrossRef]
42. Huo, A.; Zhao, Z.; Luo, P.; Zheng, C.; Peng, J.; Abuarab, M.E.L.S. Assessment of Spatial Heterogeneity of Soil Moisture in the Critical Zone of Gully Consolidation and Highland Protection. *Water* **2022**, *14*, 3674. [CrossRef]
43. Baveye, P.; Sposito, G. The Operational Significance of the Continuum Hypothesis in the Theory of Water Movement through Soils and Aquifers. *Water Resour Res.* **1984**, *20*, 521–530. [CrossRef]

44. Quintard, M.; Kaviany, M.; Whitaker, S. Two-medium treatment of heat transfer in porous media: Numerical results for effective properties. *Adv. Water Resour.* **1997**, *20*, 77–94. [CrossRef]
45. Quintard, M.; Whitaker, S. Convection, dispersion, and interfacial transport of contaminants: Homogeneous porous media. *Adv Water Resour.* **1994**, *17*, 221–239. [CrossRef]

sustainability

MDPI

Review

A Critical Analysis of Geological Hazard Risk Assessment Including Future Perspectives

Yuxiang Cheng [1,2], Yue Gan [1,*], Chunmei Shi [3], Aidi Huo [4,*], Yinghui Pei [5], Yu Song [6], Xing Wang [4] and Adnan Ahmed [4]

1 College of Geological Engineering and Geomatics, Chang'an University, Xi'an 710054, China; chengyx@chd.edu.cn
2 Key Laboratory of Western China's Mineral Resources and Geological Engineering, Ministry of Education, Chang'an University, Xi'an 710054, China
3 Shandong Jiantou Engineering Inspecting & Appraisal Co., Ltd., Qingdao 266112, China; zh431914@126.com
4 School of Water & Environmental, Chang'an University, Xi'an 710054, China; wangxing199910@163.com (X.W.); 2021129072@chd.edu.cn (A.A.)
5 Anhui Urban Construction Design Institute Co., Ltd., Hefei 230051, China; foreverphy@126.com
6 Shanxi Engineering Survey Research Institute Co., Ltd., Xi'an 710054, China; 15029576988@163.com
* Correspondence: ganyue000124@foxmail.com (Y.G.); huoaidi@chd.edu.cn (A.H.)

Abstract: Geological hazards are widely distributed, cause huge losses, and have always been the focus of attention for engineering and environmental geologists. Geological hazard evaluation is the basis of research and has important theoretical significance for preventing and controlling geological hazards. Therefore, geological hazard evaluation has become the focus of engineering and environmental geology. The question of how to build a universal index system model of geological hazard evaluation is an urgent problem that needs to be solved in geological hazard evaluation. Based on a large amount of previous research data, this paper takes landslide hazard as an example and systematically expounds the main problems that need to be solved in the current geological hazard evaluation from five aspects: basic concept, evaluation scope and accuracy, evaluation index system and evaluation criteria, evaluation method, and applicability of evaluation results. A landslide hazard assessment index system model is proposed, which applies to all regions, including all of the factors that may affect the formation of landslides. It is also hoped that this will be used as an example to establish various types of disaster evaluation and assessment systems. If the parameter has no value in the assessment process, it can be processed as 0. On this basis, further research is suggested from the perspectives of the geological hazard evaluation level, geological hazard evaluation theory, and method. To provide thoughts on and suggestions for geological hazard risk assessment method research, standard revision, investigation and evaluation, and risk management and control need to be considered.

Keywords: geological hazards; evaluation; research status; existing problems; prospects

1. Introduction

Geological hazards are combinations of geological events that damage human lives, property, and the natural environment. They are produced by the earth's natural internal and external dynamic actions or by the human-induced environmental degradation [1].

The frequent occurrence of geological hazards has caused great harm to social and economic activities and human life [2,3]. With the development of society, the issue of environmental sustainability has attracted more and more attention [4,5]. The question of how to prevent and reduce the losses caused by geological hazards has become a primary issue. Therefore, geological hazard evaluation has become a necessary research topic.

Since the United Nations published natural disaster hazard assessment methods and put forward and completed the 10-year plan (IDNDR) to reduce losses caused by geological hazards by 30%, many research results on geological hazard evaluation have emerged [6]. These include the study of geological hazard control and regionalization [7], the study of geological hazard evaluation and evaluation methods, and the construction of a geological hazard information system [8]. However, this topic can generally be divided into geohazard susceptibility, risk, and hazard evaluation [9].

In the 1970s, numerous scholars extensively employed qualitative methods to carry out a susceptibility assessment of geological hazards [10]. However, with the emergence and development of GIS technology, the challenge of conducting a superimposed analysis of geological hazard susceptibility data has been overcome, leading to a shift towards utilizing mathematical models in conjunction with GIS tools as the mainstream approach for assessing geological hazard susceptibility [11,12]. A geological hazard risk assessment is distinct from a susceptibility assessment as it incorporates temporal considerations. Hence, it is often defined as evaluating the likelihood of potential destructive phenomena occurring within a specific timeframe in a study area [13]. Assessing geological disaster hazards involves evaluating potential risks to human life, property loss, and environmental damage based on risk results; thus, vulnerability is considered alongside risk assessment [14]. Nevertheless, scholars may have varying interpretations regarding the definitions related to geological hazard susceptibility, risk, and hazard, which will be elaborated upon in Section 2.1.

This paper analyzes the existing research results, and it focuses on landslide hazards, summarizes the existing problems in the current research, and presents a preliminary idea for future geological hazard evaluation research.

2. Problems in Current Geological Hazard Evaluation Research

Upon evaluating the geological hazard evaluation research findings, it becomes evident that while some progress has been made, there are also significant prevailing issues. This article will focus on five aspects.

2.1. Basic Concept

Since the beginning of geological hazard research, problems have existed, such as the ambiguity of basic concepts and the lack of unified definitions. The connotation and extension of the concepts of geological hazards and hidden dangers of geological hazards are not clear enough. Different scholars have different understandings of these concepts. The content and boundaries of geological hazard susceptibility, risk, and hazard assessment are unclear.

Zhang [15] considered generalized geological hazards to occur due to geological (natural, artificial, or integrated) mutations of the geological environment or progressive damage and loss caused by the phenomenon of human life and property or events. Geological hazards, in a narrow sense, refer to hazards related to geological action that are caused by natural factors or human activities and endanger the safety of lives and property, including rock falls, landslides, mudslides, ground subsidence, ground collapses, ground fissures, etc. [16]. Zhang et al. [9] believe that a geological hazard is a process or phenomenon in which an inevitable geological process deteriorates the geological natural environment, destroying human life and property or seriously destroying the resources and environment that humans rely on for survival and development. A geological event destroys human life, property, and living environments. In the world, geological hazards mainly refer to collapses, landslides, and debris flows, broadly equivalent to landslides [17].

The definition of a geological hazard susceptibility assessment is relatively straightforward. Landslide susceptibility in international standards refers to the qualitative or quantitative evaluation of the type, volume (or area), and spatial distribution of existing or potential landslides in an area. Geological disaster-prone areas in China refer to regions that are prone to geological hazards. Existing and potential geological hazards should be considered in assessing susceptibility, but the time dimension should not be considered [18]. It mainly uses the principle of discussing the future with the present to analyze the situation of hazards in a specific area and predict the possibility of destructive geological hazards in the future [19]. The critical content of geological hazard susceptibility assessment is the distribution density, characteristics, and susceptibility conditions of geological hazards. Many scholars have conducted much work in this area and made some progress [20].

However, the definitions of a geological hazard risk and hazard assessment are very confusing. Some scholars believe that the risk assessment of geological hazards mainly refers to evaluating the possibility of potential destructive geological hazards in the study area over a period, which is a qualitative evaluation [20]. Other scholars believe that the risk assessment of geological hazards is a semi-quantitative assessment of the possibility of life casualties and property losses caused by geological hazards. As for the definition of hazard, different scholars have different understandings [20]. Varnes et al. [21] believe that the term hazard refers to the expected value of the number of deaths caused by a geological event at a certain period. Fell et al. [22] believe that hazard is the probability of a disaster causing losses to human life, property, and the environment. Huang and Xiang [23] believe that hazard refers to the likelihood of a disaster in the event of casualties and property losses. Zhu et al. [24] believe that geological hazards may cause the loss of life, property, and social economy within specific geographical and time ranges.

Due to the lack of clear basic concepts in the field of geological hazard assessment, the authors consulted many studies in the literature and proposed the following definitions:

Geological hazards refer to destructive geological phenomena that have occurred in the past and caused the loss of human lives and property. The core is that they have rendered human life casualties and property losses, meaning they are harmful.

Potential geological hazards refer to destructive geological phenomena that will likely undergo obvious deformation and damage within a specific evaluation period and threaten human engineering activities or human life and property. There are three situations as follows: ① existing harmful geological phenomena that may threaten human engineering activities or human life and property in the future; ② currently, they do not exist, but they can be expected to cause obvious deformation and damage in the future, which will affect human engineering activities or threaten human life and property, such as current unstable slopes that may develop into collapses or landslides in the future; and ③ bad geological phenomena that have occurred in the past and caused human life casualties and property losses (geological hazards referred to above), which may reactivate in the future and threaten human engineering activities or human life and property. There are two conditions that can determine whether adverse geological phenomena constitute hidden dangers of geological hazards. One is whether deformation and destruction will occur within a specific evaluation period, and the other is whether it will cause human life casualties and property losses after deformation and destruction. The two conditions are indispensable. Suppose deformation and destruction occur within a specific evaluation period but do not cause human life casualties and property losses. In that case, they are still classified as destructive geological phenomena.

According to the above concepts, the objects of geological hazard susceptibility evaluation are unfavorable geological phenomena and geological hazards, and the risk and hazard evaluation of geological hazards mainly target the hidden danger points of geological hazards. The assessment of geological hazard susceptibility is mature enough to allow for the use of the existing definition. Geological hazard susceptibility refers to an area that is prone to geological hazards. Adverse geological phenomena and hazards should be considered in evaluating susceptibility, but the time scale should not be considered [18]. The key content of a geological hazard susceptibility evaluation is the formation conditions and triggering factors of geological hazards. Existing adverse geological phenomena or the development status of geological hazards should not be used as indicators for evaluating the susceptibility of geological hazards but as essential bases for inspecting the results of assessing the susceptibility of geological hazards.

One object of a geological hazard risk assessment is a potential geological hazard, regardless of the time scale. One can refer to the "geological hazard risk assessment specifications" and mainly predict and evaluate the severity of the maximum human life casualties and property losses that potential geological hazards may cause. The core content predicts and evaluates potential geological hazards, including the maximum impact range, population, and total asset survey.

Another object of a geological hazard assessment may be the hidden danger point of a geological hazard, and the time scale must be considered to predict and evaluate a certain number of casualties and property losses caused by hidden danger points of geological hazards at different time scales. The core content includes the calculation of the probability of damage being caused by potential geological hazards at different time scales, predicting the impact range of geological hazards, and evaluating certain levels of casualties and property losses caused by geological hazards.

2.2. Evaluation Range and Accuracy

When analyzing the research results of a geological hazard evaluation, the scope of the assessment varies greatly, ranging from tens of square kilometers to hundreds of thousands of square kilometers [25].

Ji et al. [26] conducted a small regional geological hazard assessment in Hancheng, Shaanxi Province; Dou et al. [27] used the RF and DT models to evaluate the susceptibility of geological hazards in Izu-Oshima, Japan. The above are small-scale geological hazard assessment studies. The medium range mainly refers to the range above the county level, or the research focus. Wang and Yi [28] used AHP to divide Mianyang into high-prone areas, medium-prone areas, and low-prone areas for geological hazards; Peng et al. [29] used the information volume method to assess the susceptibility of geological hazards. Research on large-scale geological hazards mainly focuses on compiling geological hazard distribution maps above the provincial level. Aleotti and Chowdhury [30] compiled a 1:25,000 geological hazard risk zoning map using data on landslide events in Italy.

There is no mature prediction and evaluation standard for the region. Concerning the landslide hazard zoning guidelines proposed by the Australian Geomechanics Society [31], Wu et al. [20] initially proposed the application scope of the results of geological hazard zoning. According to the current progress of the national geological hazard survey, it is not appropriate to emphasize the survey distribution and susceptibility evaluation zoning map of the federal or provincial scale with a ratio of less than 1:200,000. Quality and accuracy are difficult to guarantee for such a small-scale national geological hazard risk and zoning map. The 1:50,000 scale survey and the primary- to intermediate-level dangerous zoning mapping are more suitable for land use, hazard prevention, and mitigation planning at the city and regional levels. In contrast, large-scale geological hazard surveys, reconnaissance, catalogs, and hazard and risk zoning mapping only apply to key towns and major national engineering sites.

Due to the significant difference in the scope of the evaluation, the scale of the drawing will inevitably be very different, and the accuracy of the assessment is naturally impossible to compare. Therefore, to solve the problem of an uneven evaluation accuracy, this paper recommends the method proposed by Wu [20]: the regional geological hazard evaluation should obtain the scale evaluation results of 1:200,000 and 1:50,000 as much as possible. Moreover, to solve the problem of incoherence between regions due to different scales or different evaluation methods, it is suggested that a higher administrative organ should organize the geological hazard evaluation study with the basin or geomorphic unit as the region.

2.3. Evaluation Index System and Evaluation Criteria

There are many types of geological hazards. Geological hazards include two types, sudden and slow-change types, with eight sub-categories, for a total of thirty-four types.

Some scholars mainly conduct evaluation research on a certain kind of geological hazard; Zhou and Ning [32] discussed the existing problems of the single-channel debris flow risk assessment model in terms of parameter selection, unbalanced sample data, generalization ability, and spatial variability of the debris flow system. Chen and Li [33] compared and analyzed the performance of various combinations in landslide sensitivity modeling. Yi et al. [34] used the weight factor model to analyze and evaluate the risk of land subsidence. Fan et al. [35] monitored the land subsidence in the main urban area of Nanjing and revealed the spatial distribution characteristics of land subsidence in the study area.

However, most scholars use all of the geological hazards in the evaluation area to conduct evaluation research. For example, Wen et al. [36] used the AHP attribute identification model based on the GIS platform to assess geological hazards in the study area. Chen et al. [37] used the information volume model to investigate the susceptibility of the study area to geological hazards.

The geological conditions and triggering factors of each geological disaster are different, and the forms of damage to the hazard-bearing body are also entirely different. For example, landslide disasters often damage the structures of landslides in the form of overall damage with the disintegration of landslides. The sliding body destroys the bearing body at the front edge of the landslide. The destruction of the disaster-bearing body by the debris flow includes submergence, overflow, impact, abrasion, and much more. The damage caused by ground fissures to the bearing body mainly includes breaking and dislocation. Differences in the form of damage can also lead to differences in the degree of damage. Therefore, if the same evaluation index system and standard are used for evaluation, one will inevitably lose sight of the other.

To solve this problem, we suggest that different evaluation criteria should be established according to different types of geological hazards, and the evaluation work of different types of geological hazards should be carried out, respectively. Finally, the obtained results should be superimposed to obtain the evaluation results of geological hazards.

Constructing a reasonable evaluation index system is critical in geological hazard evaluation research. This paper studies the evaluation index system used in previous studies in the literature and finds differences [27,38,39]. So far, there has neither been a standard guideline nor a unified view of the world in geological hazard assessment research. The principles, selection, quantification and normalization, and determination of weights for constructing the index system are also different.

Chau et al. [40] believe that if the relationship between influencing factors and landslide occurrence is clear and easy to quantify, it can be used as an evaluation index. Aleotti and Chowdhury [30] use expert experience to select evaluation indicators. Baeza and Corominas [41] believe that the indicators are often chosen not because they are the most suitable, but because these data are relatively easy to obtain under existing conditions. Li et al. [42] used the national geological disaster survey results to establish a four-level comprehensive evaluation index system. Wang et al. [12] built an evaluation index system based on previous research results and combined it with existing geological conditions. Hamza and Raghuvanshi [43] summarized previous research results and selected six indicators, including lithology, elevation, and aspect, to construct an index system.

In the past 30 years, although scholars worldwide have researched geological hazard evaluation and obtained many research results, due to the regional differences of geological hazards, the evaluation work lacks uniform or widely recognized evaluation standards, norms, or guidelines. Only the latest geological hazard risk assessment specifications issued by the Ministry of Land and Resources, People's Republic of China, are relatively uniform [44].

In terms of geological hazard susceptibility evaluation research, most scholars establish an evaluation index system based on the geological and environmental conditions of the evaluation area and the influencing factors of geological hazards, refer to the qualitative evaluation results, and determine the degree of susceptibility based on the differences in the results of the evaluation area. The undesirable consequence is that the two adjacent evaluation areas do not converge on the borders. At the same time, the evaluation results of different regions lack horizontal comparability. The fundamental reason for this is that the evaluation index system is not uniform, and the index quantification, normalization methods, and weights are inconsistent.

Therefore, the author suggests that some experts who are deeply engaged in the field of geological hazard evaluation and evaluation can take the lead and formulate a unified geological hazard evaluation index system and standards with the strength of the government and academia to form a systematic work and research method in this field.

2.4. Evaluation Method

Early assessments of geological hazards were based on qualitative methods, such as expert scoring methods. This type of evaluation method mainly relies on expert knowledge and experience and lacks objectivity. Since the 1980s, more and more mathematical methods have been introduced into the field of geological hazard evaluation, opening the era of the quantitative assessment of geological hazard evaluation.

Based on the research results of existing geological hazard risk assessment methods, the geological hazard risk zoning evaluation model is divided into geographic information systems, qualitative, quantitative, machine learning, coupled methods, and others. The geographic information system is mainly based on the ArcGIS platform for risk regionalization evaluation. Qualitative models are generally based on expert knowledge and experience, including expert scoring, analytic hierarchy, and weighted linear combination methods. Quantitative models mainly include statistical analysis models, including the Newmark displacement model, contribution weight overlapping addition, fuzzy comprehensive evaluation method, rough set theory, etc. Rough set theory is a small sample method that quantitatively analyses and deals with inaccurate, inconsistent, and incomplete information and knowledge. Statistical analysis models can be divided into binary statistical analysis models and multivariate statistical analysis models, where binary statistical analysis models mainly include the deterministic coefficient method, evidence weight method, information quantity method, frequency ratio method, etc. Multivariate statistical analysis models mainly include discriminant analysis, logistic regression analysis, etc. Machine learning models mainly include neural networks, random forests, support vector machines, Bayesian networks, decision trees, etc. The coupling model is mainly an evaluation model coupled with two or more methods. Other methods include the comprehensive index method, mutation theory, matter–element model, fractal theory, and quantitative theory.

Aleotti and Chowdhury [30] used the expert scoring method to quantify different control factors to evaluate the susceptibility of geological hazards. Xiang and Huang [45], Cheng et al. [46], Liu et al. [47], and Pradhan and Lee [48] used artificial neural network methods to assess geological hazards. Wang and Yi [28], P. Kayastha et al. [49], Gao and Su [50], Wang et al. [51], Chen et al. [52], and Ji et al. [26] used an analytic hierarchy process to determine the weight of evaluation indicators. Li et al. [53], Azimi et al. [54], and Fatemi Aghda et al. [55] used fuzzy logic methods to analyze the susceptibility, hazard, and risk of geological hazards. Mathematical statistics include regression models, binary statistical analysis, principal components, and discriminant analysis. Liu et al. [47] and Bai et al. [11] used a binary logistic regression model to study the relationship between geological hazards and their controlling factors; Hamza and Raghuvanshi [43] carried out a geohazard risk assessment for the Jeldu District using binary statistical analysis; Carrara et al. [56] and Baeza et al. [41] used a stepwise discriminant analysis to conduct geological hazard evaluation research.

Although many achievements have been made in the single model for geological hazard risk zoning and evaluation, there are still many places that need to be improved: the qualitative model relies too much on the subjective experience and analysis of the expert, resulting in low reliability. The quantitative model only performs a linear superposition and summation of the information entropy of the risk index, and it is difficult to determine the relationship between each factor and the disaster point in the high-dimensional space. Constrained by convergence and the number of hidden nodes, machine learning models are prone to overfitting, resulting in model distortion. At present, the coupling of two kinds of models is the most widely used method in the evaluation of geological hazard risk regionalization, which mainly includes the coupling of two kinds of quantitative models, such as the coupling of a qualitative model and quantitative model, the coupling of a quantitative model and machine learning model, and the coupling of two kinds of machine learning models. The comparison between the two models and the single model is as follows: deterministic coefficient logistic regression coupling model > deterministic coefficient > logistic regression model; information quantity–neural network coupling model > information quantity model; Newmark model–neural network coupling model > neural network model coupling > Newmark model. It can be seen that the coupled model has higher precision and better stability than the single model. In the coupling model analysis and comparison, the quantitative model is coupled with the qualitative model, the quantitative model, and the machine learning model, respectively, and the coupling accuracy and stability of the two quantitative models are obtained.

2.5. The Applicability of the Evaluation Results

There have been many studies on geological hazard assessment, but few evaluation results have been used in geological hazard prevention [57–60]. The assessment results of the susceptibility of geological hazards divide the assessment areas into high-prone, medium-prone, low-prone, and non-prone areas. According to the "Geological Disaster Regulations" requirements, urban planning and engineering construction in areas prone to geological hazards must specialize in geological hazard risk assessment. At the same time, the evaluation results of the susceptibility of geological hazards are also essential bases for local governments at all levels to carry out the prevention and control of geological hazards, as well as one of the critical decision-making bases for activities such as social and economic development planning, urban planning, and site selection for major projects. The results of the geological hazard risk assessment have few practical uses. The content of practical application value is based on the threat objects of each determined geological hazard point, and the dangerous geological hazard points can be screened out, which is beneficial to local governments or enterprises to carry out targeted geological hazard prevention and control work. However, most geological hazard risk assessments are currently in the exploratory stage, and very few research results have been genuinely applied to preventing and mitigating geological hazards.

3. Discussion and Suggestion on Geological Hazard Evaluation

Since the 21st century, geological hazard assessments have become effective means of disaster prevention and reduction, and there are abundant research materials worldwide [8,22,37,61]. They provide rich and solid primary data for evaluating and researching geological hazards. Therefore, further enhancing the depth of the evaluation and research of geological hazards is possible. Based on the problems existing in the assessment of geological hazards at this stage, the author believes that further research can be carried out using the below aspects.

3.1. Divide the Levels of Geological Hazard Evaluation and Clarify the Content of Geological Hazard Evaluation

First of all, the susceptibility of geological hazards involves a regional evaluation. However, the practicability of geological hazards and risk regional assessment results is not strong at present [40,62,63], so it can be transformed into the hazard and risk assessment of a single hidden point of geological hazards, which can save a lot of evaluation work and overcome the disadvantage of the insufficient practicability of the evaluation results. The evaluation results of the susceptibility of geological hazards can be used as an essential basis for the prevention and control of geological hazards in the process of local government socio-economic development planning, town planning, important village planning, and significant project site selection so that the construction of critical economic zones, towns, important villages, and major projects can be as far from geological hazard-prone areas as possible, even if it is impossible to prevent these events from happening [64]. We can also take targeted measures to control geological hazards [65,66]. Through the risk assessment of a single geological hazard hidden danger point, the geological hazard hidden danger points that threaten the most people and property can be determined, which would be conducive to the targeted work of the local government, especially in the preparation of geological hazard prevention and control plans. Humanpower, material resources, and financial resources are used to prevent and control individual geological hazards that threaten a large number of people and property. The results of a risk assessment have important guiding significance for the classification of geological hazards in the planning of geological hazard prevention and control. A geological hazard assessment can draw conclusions about which geological hazard hidden danger points cause a certain number of human casualties and property losses within a certain evaluation time limit. If the probability of exceeding within a short time limit is greater, the urgency of prevention is self-evident. Therefore, the hazard assessment results have important guiding significance for the stage of geological hazard prevention and control in the geological hazard prevention and control plan.

3.2. Further Research on the Theory and Method of Geological Hazard Evaluation

3.2.1. Research on the Theory and Method of Landslide Hazard Susceptibility Evaluation

We must first determine their causal relationship and existing geological hazards to assess the susceptibility of geological hazards. A lot of evaluation work uses the density of geological hazard points as one of the crucial indicators of susceptibility evaluation, and the corresponding weight is also relatively large [67]. This result is often in good agreement with the law of geological hazard development. Nevertheless, from the concept of susceptibility, the susceptibility of geological hazards is determined by regional and environmental conditions and the triggering or inducing factors of geological hazards. The high-prone areas of geological hazards are inevitably the areas where geological hazards frequently occur, so detailed survey results of geological hazards can be used to verify the evaluation results. For example, Wang et al. [68] provided evidence of the increased concentration of disaster locations in areas with high susceptibility, indicating that their coupled model evaluation of landslip susceptibility is more accurate. The causal connection between the two phenomena lies in the fact that geological risks have an increased tendency to manifest in vulnerable regions, resulting in an increased frequency of geological disasters. The increased prevalence is both the consequence and the catalyst of the elevated vulnerability.

Secondly, it is necessary to adjust the geological hazard assessment area and scope. Most existing geological hazard susceptibility evaluation studies are based on administrative units [69–71]. The main reason for this is that the detailed survey of geological hazards is carried out in county-level administrative regions. However, an administrative region often spans several geomorphic units and involves different watersheds. Different survey units, evaluation index systems, and evaluation results have caused the evaluation results of administrative areas to be poorly connected. Even with the same or similar geological conditions, environmental conditions may be at different levels of geological hazard susceptibility. Some scholars have divided the study area based on a river basin or water system unit, conducted a susceptibility evaluation, and found that the results are more consistent with the disaster distribution law [72,73]. Given this, based on the existing detailed survey data of geological hazards, a study on the susceptibility of geological hazards based on the basin should be carried out.

Third, the susceptibility of geological hazards is controlled by the conditions and triggering factors of geological hazards [38]. However, the formation conditions and triggering factors of different geological hazards are different and may even be far apart. Therefore, conducting a susceptibility assessment on all geological hazards has theoretical drawbacks. Thus, the susceptibility evaluations of slope geological hazards, debris flow hazards, ground fissure hazards, ground subsidence hazards, and ground collapse hazards can be carried out separately. Then, all of the evaluation results can be superimposed in a topological space, and finally, the zoning map of the geological hazard susceptibility of the evaluation area can be obtained.

Fourth, to make the evaluation results of geological hazard susceptibility in different regions comparable, the evaluation index system of geological hazard susceptibility and the index value and normalization rules should be unified [74]. Taking the basin as the benchmark to evaluate the area and conducting the evaluation according to the types of geological hazards makes it possible to establish this set of rules. For example, the below figure shows the basic framework (Figure 1) used to establish a slope-type geological hazard susceptibility evaluation index system and refine the evaluation factors. Then, we will study the value rules and normalization rules of different evaluation factors. At the same time, it is necessary to study the degree of influence of varying evaluation factors on the susceptibility of geological hazards and determine the unified weight. Suppose this set of rules can be established. In that case, there are norms to check for index selection and rules to follow in the value and normalization of evaluation factors, and the weights can be uniform so that the evaluation results of different evaluation basins will be comparable. The evaluation results will not have a disconnection problem between the division boundaries.

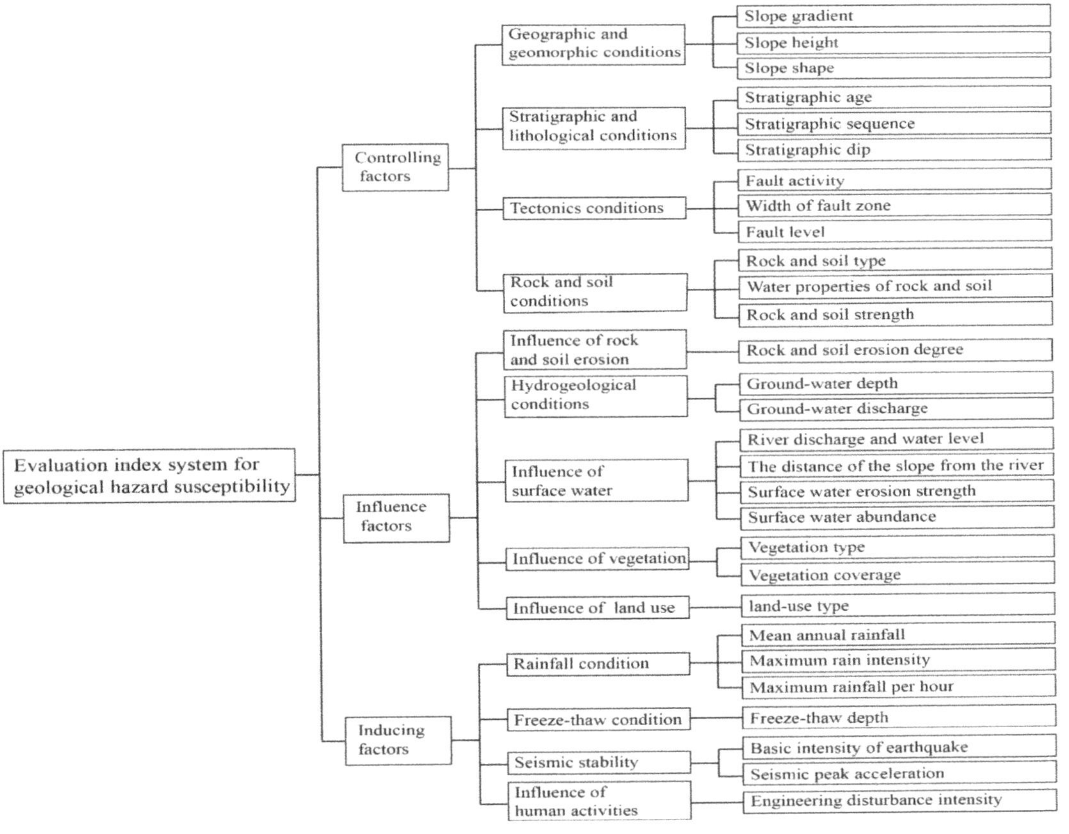

Figure 1. Susceptibility evaluation of slope geological hazards.

3.2.2. Research on Theories and Methods of Geological Hazard Danger Assessment

According to the danger assessment concept suggested above, the main goal of conducting a geological hazard danger assessment is to predict the maximum impact area of hidden danger points of geological hazards and to investigate and count the population and total assets within the impact area. Geological hazard assessments have reached a relatively mature state, so their standard formulation can be modified based on the geological hazard assessment standard issued by the Ministry of Land and Resources of China in 2015 (Table 1) [75].

Table 1. Specifications for risk assessment of geological hazards. (**a**) Geological hazard assessment grading table. (**b**) Classification of geological hazard degree. (**c**) Geological hazard classification table.

(**a**)

Importance of Construction Project	Complexity of Geological Environmental Conditions		
	Complex	Medium	Simple
Important	level one	level one	level two
Secondary importance	level one	level two	level three
Generally important	level two	level three	level three

(**b**)

Degree of Hazard	Disaster Situation		Dangerous Situation	
	Death Toll/P	Direct Economic Loss/10K	Number of People under Threat/P	Possible Direct Economic Loss/10K
Large	≥10	≥500	≥100	≥500
Medium	>3~<10	>100~<500	>10~<100	>100~<500
Small	≤3	≤100	≤10	≤100

(**c**)

Degree of Hazard	Degree of Development		
	Strong	Medium	Week
Large	High risk	High risk	Medium risk
Medium	High risk	Medium risk	Medium risk
Small	Medium risk	Low risk	Low risk

Note 1: Disaster situation refers to the geological disaster that has occurred. The index of "casualties" and "direct economic losses" is used for the evaluation. Note 2: Dangerous situation refers to the geological disaster that may occur, and the index of "number of people under threat" and "possible direct economic loss" is used for the evaluation. Note 3: The hazard degree is evaluated by the index of "disaster situation" or "dangerous situation".

3.2.3. Research on Theories and Methods of Conducting Geological Hazard Assessment

Geological hazard assessment research is weak. The authors think that we can start from the following aspects in the future.

First, the time limit for hazard assessment should be determined, which should be aimed at applying hazard assessment results. As for the government, the department carries out geological hazard management. It can be divided into short-term (one year), medium-term (five years), and long-term (ten years), corresponding to the exceedance probability of one year, five years, and ten years. The one-year probability of exceeding is mainly used to prepare the annual geological hazard prevention and control plan, and the five-year probability of exceeding is mainly used to prepare the geological hazard prevention plan. Ten-year exceedance probability is primarily used for the long-term prevention of geological hazards.

Second, prediction and evaluation research on the probability of geological hazards should be conducted in different time frames. A specific influencing factor often triggers the hidden dangers of sudden geological hazards and their occurrence, and the recurrence probability of this triggering factor is the probability of occurrence of geological hazards. Of course, the recurrence probability of the triggering factor also requires a lot of investigation, monitoring, statistics, and experimental research—for example, one of the crucial factors affecting the formation of landslides during rainfall. If we can find a correlation between rainfall or rainfall intensity and the stability of landslides, we can obtain the rainfall or rainfall intensity thresholds at which landslides occur. In this way, we can evaluate the probability of landslide failure according to the likelihood of critical rainfall or rainfall intensity threshold. The development speed can be considered uniform regarding the hidden dangers of slow-changing geological hazards. The monitoring data can be used to predict and evaluate whether casualties and property losses will occur within a specific evaluation period. Accordingly, they can be used to predict and estimate the exceedance probability of a certain magnitude of casualties and property losses within a particular evaluation period.

Third, deep machine learning can be used to identify hidden geological hazards automatically. With the rapid development of artificial intelligence technology, deep machine learning will become a tool for identifying geological hazards and developing them toward intelligent automation. The existing research results show that the new geological hazards can be interpreted automatically and quickly through remote sensing change detection, artificial intelligence, and deep machine learning because their spectral and texture characteristics significantly differ from those of the surrounding environment. Deep machine learning methods based on convolutional neural networks can effectively perform automatic recognition for ancient landslide bodies with unclear deformation, geomorphic features, and general geological hazard points. However, the current recognition accuracy is low, and further research is needed.

Fourth, surveys and statistics on the populations and properties within different impact areas of hidden danger points of geological hazards should be conducted, and the population and property loss classifications should be determined. The vulnerability assessment research of the bearing body of the hidden danger point of geological hazards should be carried out. The population and total assets specified in the geological hazard risk evaluation are the most destructive evaluations after the geological hazard occurs, and the risk level can be obtained. However, people are aware of and able to actively avoid geological hazards, and some buildings can also resist geological hazards. Therefore, it is necessary to research the vulnerability of the population and buildings to determine the classifications of population and property losses.

Finally, research should be carried out on the hazard assessment model of geological hazards. There are many geological hazard assessment model results, but they are not based on vulnerability classification and different time scales. Therefore, based on the existing research results, the occurrence probability of geological disasters with unavoidable casualties and property losses can be calculated under different time scales and vulnerability classification conditions. The shorter the time limit, the greater the probability of surpassing it, the greater the necessity of geological hazard prevention, and the more urgent the time.

4. Conclusions

(1) This paper points out the problems that need to be solved by a geological hazard assessment, including the basic concepts, evaluation scope and accuracy, evaluation index systems and criteria, evaluation methods, and the applicability of the evaluation results.

(2) At present, the risk assessment of geological hazards in relevant studies in the literature mainly focuses on the possibility (i.e., susceptibility and risk) assessment of geological hazards, and there is insufficient research on the classification of geological hazard assessment levels and the theory and method of geological hazard assessment.

(3) In the past, evaluation indicators were often chosen due to different regions, types, influencing factors, and even different levels of data integrity. Each project established its own evaluation index system. Therefore, we proposed a landslide hazard assessment index system model that is applicable to all regions, which contains all of the factors that may affect the formation of landslide hazards. If a parameter with no value is encountered during the evaluation process, it only needs to be processed as 0 to be included in the calculation.

A geological hazard assessment is the basis for the risk control and prevention of geological disasters. Although many scholars have made considerable research and achievements in geological disaster assessment and evaluation, due to the complexity of this field, there are still problems, such as inconsistent evaluation standards, unpersuasive evaluation methods, and the insufficient practicability of the evaluation results. Therefore, geological disaster assessment research should be carried out. There is still a long way to go to obtain reasonable and practical geological hazard evaluation results. The authors put forward some understanding of this, hoping to play a role in promoting the rapid development of geological hazard evaluation research. The following are suggestions for further steps:

(1) A universal evaluation index system and index assignment rules should be constructed. It is recommended to consider the control factors that contribute to the formation of geological hazards, including the topography, stratigraphic lithology, geological structure, hydrogeological conditions, geological conditions of rock and soil engineering, and influencing factors of geological disasters, including rainfalls, earthquakes, human activities, etc. A geological hazard evaluation index system applicable to all regions should be established. According to the value range of the geological hazard index, universally applicable index assignment rules should be established. The evaluation result value interval should be studied and the unified classification threshold should be delimited. This can solve the problem of the lack of comparability of geological hazard evaluation results.

(2) The accuracy of geological hazard investigation should be improved, and a connection between geological hazard risk assessment results and geological hazard prevention and control should be formed. Geological hazard investigations and evaluations at scales of 1:100,000 and 1:50,000 have been completed. Regional geological hazard investigations and evaluations at a scale of 1:10,000 should be carried out in the future, and geological hazard surveying and evaluations at a scale of 1:2000 should be carried out in critical areas. The research results can be directly applied to the design of geological hazard prevention and control and the risk management of geological hazards.

Author Contributions: All authors contributed to this study's conception and design. Y.C. first proposed the idea of the article. All the authors participated in the literature search and analysis. Y.P. wrote the first draft of the manuscript. Y.G. and Y.S. revised the article, and A.H. conducted the final review and revision. All authors commented on previous versions of the manuscript. All authors have read and agreed to the published version of the manuscript.

Funding: This research was funded by the National Natural Science Foundation of China (Grant Nos. 42377158 and 42261144749); a high-tech research cultivation project (Grant No. 300102268202); the Program for Key Science and Technology Innovation Team in Shaanxi Province (Grant No. 2014KCT-27); Promoting Higher Education Reform by Big Data Technology, Energizing Future Exploration in Teachers and Students (FPE, Grant No: 2017B00022); and the Shaanxi Provincial Institute of Higher Education 2017 Annual Higher Education Research Project (Grant No: XGH17049).

Data Availability Statement: The datasets used and analyzed in this study are available from the corresponding author upon reasonable request.

Acknowledgments: The authors are grateful to the reviewers for their insightful remarks that helped improve the manuscript. They also thank Adnan Ahmed for polishing the article.

Conflicts of Interest: Authors Chunmei Shi was employed by the company Shandong Jiantou Engineering Inspecting & Appraisal Co., Ltd. Authors Yinghui Pei was employed by the company Anhui Urban Construction Design Institute Co., Ltd. Authors Yu Song was employed by the company Shanxi Engineering Survey Research Institute Co., Ltd. The remaining authors declare that the research was conducted in the absence of any commercial or financial relationships that could be construed as a potential conflict of interest.

References

1. Luo, Y.; Zhang, L.; Zhang, Y. *Methods of Geological Hazard Risk Assessment*; Geological Publishing House: Beijing, China, 1998; ISBN 7-116-02650-9.
2. Zhao, Z.; Huo, A.; Cheng, Y.; Luo, P.; Peng, J.; Elbeltagi, A.; EL-Sayed Abuarab, M.; Mokhtar, A. Experimental Study on Slope Morphological Characteristics and Stability Analysis of GCHP Engineering in the Loess Plateau. *Adv. Space Res.* **2023**, *72*, 4324–4335. [CrossRef]
3. Huo, A.; Peng, J.; Cheng, Y.; Zheng, X.; Wen, Y. Temporal Characteristics of the Rainfall Induced Landslides in the Chinese Loess Plateau (China). In *Recent Advances in Geo-Environmental Engineering, Geomechanics and Geotechnics, and Geohazards*; Kallel, A., Erguler, Z.A., Cui, Z.-D., Karrech, A., Karakus, M., Kulatilake, P., Shukla, S.K., Eds.; Advances in Science, Technology & Innovation; Springer International Publishing: Cham, Switzerland, 2019; pp. 425–427, ISBN 978-3-030-01664-7.
4. Huo, A.; Li, H. Assessment of Climate Change Impact on the Stream-Flow in a Typical Debris Flow Watershed of Jianzhuangcuan Catchment in Shaanxi Province, China. *Environ. Earth Sci.* **2013**, *69*, 1931–1938. [CrossRef]
5. Huo, A.; Zhao, Z.; Luo, P.; Zheng, C.; Peng, J.; Abuarab, M.E.-S. Assessment of Spatial Heterogeneity of Soil Moisture in the Critical Zone of Gully Consolidation and Highland Protection. *Water* **2022**, *14*, 3674. [CrossRef]
6. UNDRO-United Nations Disaster Relief Co-Ordinator. *Mitigating Natural Disasters: Phenomena, Effects and Options: A Manual for Policy Makers and Planners*; United Nations: New York, NY, USA, 1991.
7. Zhou, P.G.; Chen, H.Q. Research on Geologic Hazard Risk Management in China Based on Geologic Hazard Survey and Zoning. *IJRAM* **2008**, *8*, 362. [CrossRef]
8. Tong, B.; Li, Y.; Yang, X.; Yin, C.; Qu, X.; Fang, H.; Han, B.; Zhang, Y. The Development of China National Geohazard Information System. *IOP Conf. Ser. Earth Environ. Sci.* **2020**, *570*, 042057. [CrossRef]
9. Zhang, L.; Zhang, Y.; Luo, Y. *Theory and Practice of Geological Disaster Assessment*; Geological Publishing House: Beijing, China, 1998.
10. Reichenbach, P.; Rossi, M.; Malamud, B.D.; Mihir, M.; Guzzetti, F. A Review of Statistically-Based Landslide Susceptibility Models. *Earth-Sci. Rev.* **2018**, *180*, 60–91. [CrossRef]
11. Bai, S.-B.; Wang, J.; Lü, G.-N.; Zhou, P.-G.; Hou, S.-S.; Xu, S.-N. GIS-Based Logistic Regression for Landslide Susceptibility Mapping of the Zhongxian Segment in the Three Gorges Area, China. *Geomorphology* **2010**, *115*, 23–31. [CrossRef]
12. Wang, N.; Shi, T.; Peng, K.; Zhang, W.; Jin, X. Assessment of Geohazard Susceptibility Based on RS and GIS Analysis in Jianshi County of the Three Gorges Reservoir, China. *Arab. J. Geosci.* **2015**, *8*, 67–86. [CrossRef]
13. Tan, S.; Zhao, X.; Li, Y.; Wei, D.; Yang, L. Risk assessment on the geological disasters based on GIS and information content model—Taking Qiubei County, Yunnan Province as an example. *J. Northwest Norm. Univ. (Nat. Sci.)* **2018**, *54*, 63–76. [CrossRef]
14. Xu, J.; Zhang, M.; Fan, W. An Overview of Geological Disaster risk Assessment. *J. Catastrophol.* **2015**, *30*, 130–134. [CrossRef]
15. Zhang, Y. On the Standardization of the Terminology Standard System for Geological Hazards. *China Qual. Stand. Guide* **2018**, 30–33. Available online: https://d.wanfangdata.com.cn/periodical/zgbzdb201805010 (accessed on 2 April 2024).
16. Department of Policies and Regulations, Ministry of Land and Resources. *Interpretation of Regulations on Prevention and Control of Geological Disasters*; Department of Policies and Regulations, Ministry of Land and Resources: Beijing, China, 2004.
17. Dai, F.C.; Lee, C.F.; Ngai, Y.Y. Landslide Risk Assessment and Management: An Overview. *Eng. Geol.* **2002**, *64*, 65–87. [CrossRef]
18. Zhang, M.; Tang, Y. Methods and Practice of Geological Hazard Risk Investigation. *Geol. Bull.* **2008**, *27*, 1205–1216. [CrossRef]
19. Rao, P.; Cao, R.; Jiang, W. Geological hazard susceptibility evaluation based on geographically weighted regression model in Yunnan Province. *J. Nat. Disasters* **2017**, *26*, 134–143. [CrossRef]
20. Wu, S.; Shi, J.; Zhang, C.; Wang, T. Preliminary discussion on technical guidelines for geological hazard risk assessment. *Geo-Log. Bull.* **2009**, *28*, 995–1005. [CrossRef]
21. Varnes, D.J.; Commission on Landslides and Other Mass Movements. The Principles and Practice of Landslide Hazard Zonation. *Bull. Int. Assoc. Eng. Geol.* **1981**, *23*, 13–14. [CrossRef]
22. Fell, R.; Corominas, J.; Bonnard, C.; Cascini, L.; Leroi, E.; Savage, W.Z. Guidelines for Landslide Susceptibility, Hazard and Risk Zoning for Land Use Planning. *Eng. Geol.* **2008**, *102*, 85–98. [CrossRef]
23. Huang, R.; Xiang, X. Regional Landslide Geological Hazard Risk Assessment and Risk Managemen. Ph.D. Thesis, Chengdu University of Technology, Chengdu, China, 2005.
24. Liangfeng, Z.; Guirong, Z.; Kunlong, Y.; Liang, Z. Risk Analysis System of Geo—Hazard Based on GIS Technique. *J. Geogr. Sci.* **2002**, *12*, 371–376. [CrossRef]
25. van Westen, C.J.; van Asch, T.W.J.; Soeters, R. Landslide Hazard and Risk Zonation—Why Is It Still so Difficult? *Bull. Eng. Geol. Environ.* **2006**, *65*, 167–184. [CrossRef]
26. Ji, Y.; Li, C.; Gao, S.; Chen, J.; Zheng, M. Risk Assessment of Geological Hazards of Hancheng City in Shaanxi Province—All Databases. *Catastrophology* **2018**, *33*, 194–200. [CrossRef]
27. Dou, J.; Yunus, A.P.; Tien Bui, D.; Merghadi, A.; Sahana, M.; Zhu, Z.; Chen, C.-W.; Khosravi, K.; Yang, Y.; Pham, B.T. As-sessment of Advanced Random Forest and Decision Tree Algorithms for Modeling Rainfall-Induced Landslide Susceptibility in the Izu-Oshima Volcanic Island, Japan. *Sci. Total Environ.* **2019**, *662*, 332–346. [CrossRef]
28. Wang, Z.; Yi, F. Assessment of the susceptibility of geological disasters in Mianyang City based on the analytic hierarchy process. *J. Nat. Disasters* **2009**, *18*, 14–23. [CrossRef]

29. Peng, K.; Peng, H.; Liang, F.; Huang, C.; Qiu, Z. Geological hazard susceptibility zoning in Ganzhou based on information model. *Saf. Environ. Eng.* **2018**, *25*, 22–28. [CrossRef]
30. Aleotti, P.; Chowdhury, R. Landslide Hazard Assessment: Summary Review and New Perspectives. *Bull. Eng. Geol. Environ.* **1999**, *58*, 21–44. [CrossRef]
31. Australian Geomechanics Society. *Landslide Risk Management*; Australian Geomechanics Society: Sydney, NSW, Australia, 2007.
32. Zhou, A.; Ning, Z. Research on related issues of single-ditch debris flow risk assessment model. *Geogr. Sci.* **2020**, *40*, 1385–1393. [CrossRef]
33. Chen, W.; Li, Y. GIS-Based Evaluation of Landslide Susceptibility Using Hybrid Computational Intelligence Models. *CATENA* **2020**, *195*, 104777. [CrossRef]
34. Yi, Y.; Liu, H.; Liu, X.; Zhang, Y.; Qi, J. Risk assessment of land subsidence under building load based on weight factor model. *Earth Environ.* **2017**, *45*, 478–489. [CrossRef]
35. Fan, X.; Li, M.; Pan, J.; Wang, S. Monitoring and Risk Assessment of Land Subsidence along Nanjing Subway. *Bull. Surv. Mapp.* **2019**, 123–126+141. Available online: https://kns.cnki.net/kcms2/article/abstract?v=29axctaKF3wS7DZn-BrmX0KW2pwb8LPmgG9vk9k9tXfOIxHwucKlXSxM4hNTTEBb9afODm_RiQHqeFAc0A3DFoc-nmaNHQRWXspqoVfWdm_gnN-plrEIUw9CmD2QqfBjiLlngDZ42Ms=&uniplatform=NZKPT&language=CHS.2024.04.27 (accessed on 2 April 2024).
36. Wen, T.; Zhou, Z.; Ying, S.; Wang, S.; Bai, S.; Zhou, J. Risk assessment of geological disaster based on GIS and AHP attribute identification model: A case study of Fuling shale gas exploitation area. *J. Chongqing Norm. Univ. (Nat. Sci. Ed.)* **2020**, *37*, 68–74+2+148. [CrossRef]
37. Chen, L.; Li, L.; Wu, F.; Xu, Y. Evaluation of geological hazard susceptibility in Beiliu City based on GIS and information method. *Earth Environ.* **2020**, *48*, 471–479. [CrossRef]
38. Jiang, H.; Yu, Y. Research on Geological Hazard Risk Assessment Method—Taking Torch Development Zone of Zhongs-Han City as an Example. *Urban Constr. Theory Res.* **2023**, 187–189. Available online: https://kns.cnki.net/kcms2/article/abstract?v=29axctaKF3ypKPZ-4TSgFTPkoWTF5aHflsXBa-6ZmHDm3aAln3tZS4RuskJUjnbsSgXx_uov3_d6qoCYBQnA-iePUWrffXGYuqkysaSVJswc4yhW1FgWMsQolr_Br3w3duMVxqCmQkM=&uniplatform=NZKPT&language=CHS (accessed on 2 April 2024).
39. Kong, J.; Zhuang, J.; Peng, J.; Zhan, J.; Ma, P.; Mu, J.; Wang, J.; Wang, S.; Zheng, J.; Fu, Y. Evaluation of Landslide Susceptibility in Chinese LoessPlateau Based on IV-RF and IV-CNN Coupling Models. *Earth Sci.* **2023**, *48*, 1711–1729. [CrossRef]
40. Chau, K.T.; Sze, Y.L.; Fung, M.K.; Wong, W.Y.; Fong, E.L.; Chan, L.C.P. Landslide Hazard Analysis for Hong Kong Using Landslide Inventory and GIS. *Comput. Geosci.* **2004**, *30*, 429–443. [CrossRef]
41. Baeza, C.; Corominas, J. Assessment of Shallow Landslide Susceptibility by Means of Multivariate Statistical Techniques. *Earth Surf. Process. Landforms* **2001**, *26*, 1251–1263. [CrossRef]
42. Li, Y.; Qu, X.; Fang, H.; Yang, X.; Yin, C. Research on comprehensive evaluation index system and evaluation method of geo-logical hazards. *Hydrogeol. Eng. Geol.* **2013**, *40*, 129–132. [CrossRef]
43. Hamza, T.; Raghuvanshi, T.K. GIS Based Landslide Hazard Evaluation and Zonation—A Case from Jeldu District, Central Ethiopia. *J. King Saud Univ.—Sci.* **2017**, *29*, 151–165. [CrossRef]
44. Geological Environment Monitoring Institute of China Geological Survey Geological Hazard Risk Assessment Specification 2021. Available online: https://std.cgs.gov.cn/content/7150668158313959424 (accessed on 2 April 2024).
45. Xiang, X.; Huang, R. Application of GIS-Based Artificial Neural Network Model in the Risk Zoning of Geological Hazards. *Chin. J. Geol. Hazard Control* **2000**, 26–30. Available online: https://kns.cnki.net/kcms2/article/abstract?v=29axctaKF3y96iskYLX82SCDdwrurEWp6PixU-2xlB1uJ5ZvZS3FabBXRYW4harveH28Mq1gOBe6WskfEHWDn3ZCmr0qpQyWITdAVav-0PPDd4ODa7gRZ9lLj06rrmN_&uniplatform=NZKPT&language=CHS (accessed on 2 April 2024).
46. Cheng, Y.; Ren, C.; Zhang, J. Discussion on Risk Assessment Method of Geological Disaster based on BP Neural Network: A Case Study of Tianshui Area. *Chin. J. Geol. Hazard Control* **2008**, 100–104. Available online: https://kns.cnki.net/kcms2/article/abstract?v=29axctaKF3zxPr75AOKUwVNfOGNOhdQ8PMDQg6-Onipi2nixL4QmX16NhKXKZdkm0LH9xNHRY9MPGRkkyr8mz_KOGcPdnUjFGkjT6PvtN6aJ3iWL2QwuS0oQwSPJOkN4&uniplatform=NZKPT&language=CHS (accessed on 2 April 2024).
47. Liu, Y.; Yin, K.; Liu, B. Application of Logistic Regression and Artificial Neural Network Model in the Spatial Prediction of Landslide Hazards. *Hydrogeol. Eng. Geol.* **2010**, *37*, 92–96. [CrossRef]
48. Pradhan, B.; Lee, S. Landslide Susceptibility Assessment and Factor Effect Analysis: Backpropagation Artificial Neural Networks and Their Comparison with Frequency Ratio and Bivariate Logistic Regression Modelling. *Environ. Model. Softw.* **2010**, *25*, 747–759. [CrossRef]
49. Kayastha, P.; Dhital, M.R.; De Smedt, F. Application of the Analytical Hierarchy Process (AHP) for Landslide Susceptibility Mapping: A Case Study from the Tinau Watershed, West Nepal. *Comput. Geosci.* **2013**, *52*, 398–408. [CrossRef]
50. Gao, L.; Su, J. A small area debris flow risk assessment method based on information entropy and AHP model. *Res. Soil Water Conserv.* **2017**, *24*, 376–380+2. [CrossRef]
51. Wang, S.; Guo, L.; Zhao, L. Risk assessment of geological disasters in Linjiang City, Jilin Province based on fuzzy analytic hierarchy process. *J. Heilongjiang Inst. Technol.* **2018**, *32*, 16–20+26. [CrossRef]
52. Chen, F.; Guo, S.; Xiong, R.; Zhong, L. Geological hazard risk assessment based on analytic hierarchy process. *Non-Ferr. Met. Sci. Eng.* **2018**, *9*, 54–60. [CrossRef]

53. Li, X.; Lu, Y.; Zhang, J.; Qi, C.; Cheng, Y. Evaluation on Urban Environment Engineering Geology by Fuzzy Logic Based on GIS—Taking Tianshui City of Gansu Province fo rExample. *J. Water Resour. Archit. Eng.* **2006**, *4*, 31–35. [CrossRef]
54. Azimi, S.R.; Nikraz, H.; Yazdani-Chamzini, A. Landslide Risk Assessment by Using a New Combination Model Based on a Fuzzy Inference System Method. *KSCE J. Civ. Eng.* **2018**, *22*, 4263–4271. [CrossRef]
55. Fatemi Aghda, S.M.; Bagheri, V.; Razifard, M. Landslide Susceptibility Mapping Using Fuzzy Logic System and Its Influences on Mainlines in Lashgarak Region, Tehran, Iran. *Geotech. Geol. Eng.* **2018**, *36*, 915–937. [CrossRef]
56. Carrara, A.; Cardinali, M.; Detti, R.; Guzzetti, F.; Pasqui, V.; Reichenbach, P. GIS Techniques and Statistical Models in Evaluating Landslide Hazard. *Earth Surf. Process. Landforms* **1991**, *16*, 427–445. [CrossRef]
57. Maffucci, R.; Ciotoli, G.; Pietrosante, A.; Cavinato, G.P.; Milli, S.; Ruggiero, L.; Sciarra, A.; Bigi, S. Geological Hazard As-sessment of the Coastal Area of Rome (Central Italy) from Multi-Source Data Integration. *Eng. Geol.* **2022**, *297*, 106527. [CrossRef]
58. Pan, Z.; Lang, Q.; Zhang, Y.; Zhang, J.; Yu, C.; Wu, C. Geological Hazard Assessment of Secondary Collapses Due to Volcanic Earthquakes on Changbai Mountain in China. *ISPRS Int. J. Geo-Inf.* **2023**, *12*, 307. [CrossRef]
59. Zou, F.; Che, E.; Long, M. Quantitative Assessment of Geological Hazard Risk with Different Hazard Indexes in Moun-tainous Areas. *J. Clean. Prod.* **2023**, *413*, 137467. [CrossRef]
60. Wang, Z.; Du, X.; Sun, Y.; Song, Y.; Dong, L.; Zhou, Q.; Jiang, W. Risk Zonation of Submarine Geological Hazards in the Chengdao Area of the Yellow River Subaqueous Delta. *Front. Mar. Sci.* **2023**, *10*, 1285437. [CrossRef]
61. Guzzetti, F.; Carrara, A.; Cardinali, M.; Reichenbach, P. Landslide Hazard Evaluation: A Review of Current Techniques and Their Application in a Multi-Scale Study, Central Italy. *Geomorphology* **1999**, *31*, 181–216. [CrossRef]
62. Catani, F.; Casagli, N.; Ermini, L.; Righini, G.; Menduni, G. Landslide Hazard and Risk Mapping at Catchment Scale in the Arno River Basin. *Landslides* **2005**, *2*, 329–342. [CrossRef]
63. Novelo-Casanova, D.A.; Oropeza, O.; Mansilla, E.; Macías, J.L.; Alcíntara, I.; Cantarero, F.J.; Figueroa, M.; Rodríguez-Van Gort, F.; Sánchez-Núñez, J.M. Integrated Risk Assessment to Natural Hazards: Case Study—Motozintla, Chiapas, Mexico. In *Disaster Management and Human Health Risk III*; WIT Press: Billerica, MA, USA, 2013; pp. 281–291.
64. Wang, C.; Wang, X.; Zhang, H.; Meng, F.; Li, X. Assessment of Environmental Geological Disaster Susceptibility under a Multimodel Comparison to Aid in the Sustainable Development of the Regional Economy. *Environ. Sci. Pollut. Res.* **2023**, *30*, 6573–6591. [CrossRef]
65. Zhao, Z.; Huo, A.; Cheng, Y.; Luo, P.; Peng, J.; Elbeltagi, A.; Abuarab, M.E.-S.; Mokhtar, A.; Ahmed, A. Impacts of Different Gully Consolidation and Highland Protection Models on the Runoff and Sediment Yield in Small Watershed of the Chinese Loess Plateau—A Case Study of Fengbugou in Qingyang City of Gansu. *Water* **2023**, *15*, 2764. [CrossRef]
66. Zhao, Z.; Huo, A.; Liu, Q.; Peng, J.; Elbeltagi, A.; Abuarab, M.E.-S.; Abu-Hashim, M.S.D. Spatiotemporal Variation in the Coupling Relationship between Human Activities and Soil Erosion—A Case Study in the Weihe River Basin. *Sustainability* **2023**, *15*, 10785. [CrossRef]
67. Wang, B. Geo-Environmental Quality Evaluation Based on GIS in Shiyan-Wudang Mountain Area. *IJEPP* **2019**, *7*, 72. [CrossRef]
68. Wang, H.; Xu, J.; Tan, S.; Zhou, J. Landslide Susceptibility Evaluation Based on a Coupled Informative–Logistic Regression Model—Shuangbai County as an Example. *Sustainability* **2023**, *15*, 12449. [CrossRef]
69. Liu, J.; Dai, B. Geological hazard risk assessment based on GIS in Gaolan County, Lanzhou City. *Resour. Inf. Eng.* **2023**, *38*, 64–68+73. [CrossRef]
70. Liu, B.; Chen, G.; Cheng, G. Risk assessment of geological disasters in Nanjing. Jiangsu Province. *Chin. J. Geol. Hazard Control* **2023**, *34*, 97–104. [CrossRef]
71. Yu, B.; Chang, M.; Ni, Z.; Sun, W.; Xu, H. Landslide Hazard Assessment in Northeast Afghanistan Plateau Based on Optimized Neural Network. *Earth Sci.* **2023**, *48*, 1825–1835. [CrossRef]
72. Hou, R.; Li, Z.; Chen, N.; Tian, S.; Liu, E.; Ni, H. Modeling of Debris Flow Susceptibility Assessment in Tianshan Based on Watershed Unit and Stacking Ensemble Algorithm. *Earth Sci.* **2023**, *48*, 1892–1907. [CrossRef]
73. Ji, W.; He, Y.; Wang, L.; Liu, S.; Chen, B. Machine Learning Solution for Landslide Susceptibility Based onHydrographic Division: Case Study of Fengjie County in Chongqing. *Earth Sci.* **2023**, *48*, 2024–2038. [CrossRef]
74. Segoni, S.; Pappafico, G.; Luti, T.; Catani, F. Landslide Susceptibility Assessment in Complex Geological Settings: Sensitivity to Geological Information and Insights on Its Parameterization. *Landslides* **2020**, *17*, 2443–2453. [CrossRef]
75. China Geological Environment Monitoring Institute Specification of Risk Assessment for Geological Hazard 2015. Available online: https://std.cgs.gov.cn/content/102 (accessed on 2 April 2024).

sustainability

MDPI

Article

Evaluation of Water Quality from the Zimny Sztok Spring (Southern Poland)—Preliminary Results

Marek Ruman and Dominika Dąbrowska *

Faculty of Natural Sciences, University of Silesia in Katowice, Bedzinska 60, 41-200 Sosnowiec, Poland; marek.ruman@us.edu.pl

* Correspondence: dominika.dabrowska@us.edu.pl; Tel.: +48-3689251

Abstract: Enabling citizens to have access to good-quality water is an important challenge in the 21st century. Water quality in springs located in urban areas may deteriorate as a result of natural and anthropogenic factors. The quality of water in springs is influenced not only by rock weathering, evapotranspiration and climate change but also by the lack of sustainable resource use. Anthropogenic activities, including industrial and domestic and agricultural activities, may not only lead to an increase in pollutants such as nitrates, chlorides and heavy metals but also to the development of dangerous bacteria. Research on water quality was conducted in two series of measurements at the Zimny Sztok spring (southern Poland), located in an urban space. The results of the analyses indicate that the waters of this spring, despite anthropogenic pressure, exceeded the permissible values for drinking water only in terms of the total number of microorganisms. Unfortunately, this was not confirmed by Water Quality Index (WQI) values. These were approximately equal to 99 in the first measurement series and about 41 in the second series. Maintaining appropriate water quality is one of the goals of rational water management, and the constant control of selected parameters in the water falls within the scope of the concept of smart city/smart water. The example of this spring shows that the water was used by residents without information about its current physicochemical parameters and bacteriological composition. This problem can be solved in the future by implementing a monitoring system for this point. The aim of this work was to examine the quality of water in the spring, use the WQI to determine the potential risks and present the general problem of people consuming water from urban sources without knowing about their parameters.

Keywords: spring; water quality; rational water management; smart city

1. Introduction

One of the most important factors influencing the quality of life is health, i.e., the state of physical, mental and social well-being. To achieve a good health state, it is necessary to meet numerous conditions, including economic and environmental factors (e.g., access to good quality water, terrain, weather conditions) [1]. Sustainable urban development is one of the most serious problems of the 21st century [2,3]. Moreover, such development would not be possible without the ability to use water resources for drinking or industrial purposes in some regions [4]. According to Sustainable Development Goal 6 (SDG 6), access to sustainable water and sanitation should be provided to all people by 2030 [5]. However, the fact is that, in many countries, there is still a lack of adequate water quality, which also leads to numerous waterborne diseases [6–8]. It is estimated that over 3 billion people do not have access to safely managed sanitation facilities [9]. Moreover, the lack of appropriate education regarding the proper management of water resources causes an increase in the amount of pollutants in water in urban spaces [10]. This problem mainly affects underdeveloped countries, where only a small percentage of wastewater is treated in centralized municipal sewage treatment plants, and the rest is discharged into water sources without any treatment, thus leading to overall resource pollution [11].

Water is crucial not only for the proper functioning of society but also for the global economy. Statistics show that only 3% of water is in its natural state [12–14]. In this context, it is extremely important to obtain data and analyze them effectively in order to protect the water [15,16].

Surface water resources are crucial for socio-economic development and the protection of ecological systems and water supply for people. However, in countries with depleted surface water resources, most drinking water comes from groundwater resources [17].

The element connecting the ecosystem of surface and groundwater is springs. Water from these springs is very often used for drinking purposes in karst areas but also in areas used for recreation [18,19]. The water component of springs is influenced by both natural and anthropogenic factors. The first group includes geological conditions, climatic conditions and land development, as well as land use [20]. Due to the fact that water in springs appears on the surface through a network of faults, cracks or contact with rocks [21], it is particularly threatened by migrating leachates [22,23]. The topography of the area also influences the amount of pollution [24]. The spatial development of the immediate vicinity of the springs, including the presence of, e.g., landfills and sewage treatment plants, definitely increases the risk of contamination [25,26]. The migration of leachates is also stimulated by the amount of precipitation. In addition to the physicochemical parameters of spring waters, an important issue when determining their suitability for consumption is the content of microorganisms [18]. Water intended for drinking should not contain pathogens [27], but spring water often contains bacteria, fungi and viruses that cause various disease epidemics [28].

Drinking water is water that is of acceptable quality (according to the WHO's guidelines or national standards for drinking water quality) in terms of physical, chemical and microbiological properties, which can therefore be used safely [29]. Due to the fact that maintaining the appropriate quality of drinking water is necessary for ensuring human health, it is essential to supervise this quality by monitoring this water through constant control using sensors. The latter are a guarantee of constant control and are included in the functioning of the 'smart city' [30]. Smart cities are an innovative concept in managing urban environments and are intended to guarantee sustainable development and improve the quality of life by combining technical, institutional and economic concepts together with modern information and communication methods. This concept is based on the use of systems that monitor environmental parameters (e.g., sensors). And this solution can be used in every place, not only in cities. One of the elements of this system is what is termed 'smart water', which is based on the use of economical sensors that can provide data through the use of cellular networks, for example. Such infrastructure is key for ensuring that the population drinks water of sufficient quantity and quality under a wide range of changing operating conditions. Thanks to the large amount of data obtained from sensors, it is easier to assess the water quality in a given area, which also allows for risk assessment of these waters. Such actions will enable the implementation of the goals of the smart water concept and the rational management of water resources. Based on the data, it is possible to conduct a water risk analysis by, e.g., using indicator methods like the Water Quality Index (WQI) [31–34].

Springs in urban spaces should be subject to water quality monitoring in order to determine changes in their physicochemical and microbiological parameters and to identify factors that may cause these changes. This is important for maintaining appropriate water quality, limiting the development of diseases caused by the consumption of water of reduced quality and implementing a sustainable water management strategy in municipalities.

One example of a location where water from a spring is drawn by residents of nearby cities and tourists is the Zimny Sztok spring in Stanica (southern Poland). Due to it being located near a road and in constant use, the spring is also at risk of pollutant migration. This research included two measurement series in 2023 and 2024 of selected physicochemical and bacteriological parameters. The aim of this paper is to examine the quality of water in the spring, determine potential risks by using the WQI and present the general problem of people consuming water from urban sources without knowing about their parameters. These are the preliminary results of this research that will be developed and are intended to illustrate the problem of the lack of monitoring of water in springs.

2. Study Area

This paper concerns a Quaternary spring located in the town of Stanica (Pilchowice commune), in the southern part of Poland (Figure 1). In terms of physicogeography, this area is part of the Silesian Upland macroregion and the Rybnik Plateau mesoregion [35]. The spring is located at an altitude of 224.6 m above sea level. The research area is located in the transitional continental–marine climate zone. The average annual temperature here is 8 °C; the average annual rainfall is about 700–800 mm. The area is dominated by southwestern and north-western winds with average speeds of 2.5 m/s, which carry dust and gas pollution to the surrounding cities. Hydrographically, the area is located in the Oder basin, in the Upper Oder water region, in the Ruda catchment area [36].

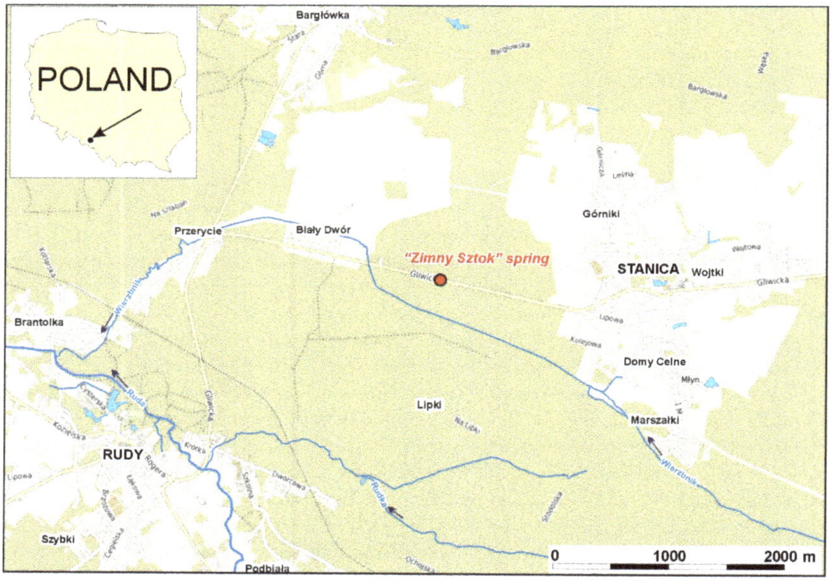

Figure 1. Study area.

The spring is located in a forest area. The distance from the source to the nearest village is 1.25 km to the east. The village of Stanica, which belongs to the Pilchowice commune, is characterized primarily by single-family buildings. About 30% of the commune is connected to the sewage system. The spring is also located 500 m west of individual single-family houses (Biały Dwór) that belong to the Rudy commune. Work is currently underway to connect these houses to the sewage network. The project of connecting houses to the sewage system for the entire Rudy commune is expected to last until 2027. The distance from the spring to the road is 7 m. There are agricultural fields approximately 400 m to the east and 700 m to the north.

The oldest geological formations in this area form a structural unit known as the Upper Silesian block. These are Precambrian crystalline rocks, mainly mica schists and paragneisses, on which there are Cambrian (siltstone and diabase), Devonian (sandstone and then limestone and dolomite with numerous inserts of claystone, mudstone and marls), Lower Carboniferous rocks (mainly sandstone, mudstone, shale, conglomerate) and carbonaceous sediment of Upper Carboniferous productive deposits. The Upper Carboniferous sediments belong to the older paralic series, which are composed of fragmentary rocks with interlayers of marine sediments, and developed in the form of gray mudstone, sandy mudstone and claystone interbedded with fine-grained sandstone and coal seams. Between the sediments of the paralic and mudstone series, there is the local belt of the Upper Silesian sandstone series in the form of medium- and fine-grained sandstone and, less often, coarse-grained sandstone and conglomerates with interbedded mudstone, among which there are thicker coal seams. The sediments of the Carpathian Foredeep include Miocene clay and sand with siderites, as well as sandy and marly clay, sand, clay shale with gypsum and anhydrite. These formations are covered by Quaternary deposits, represented mainly by sand, gravel, clay, glacial boulders and boulder clay (Figure 2).

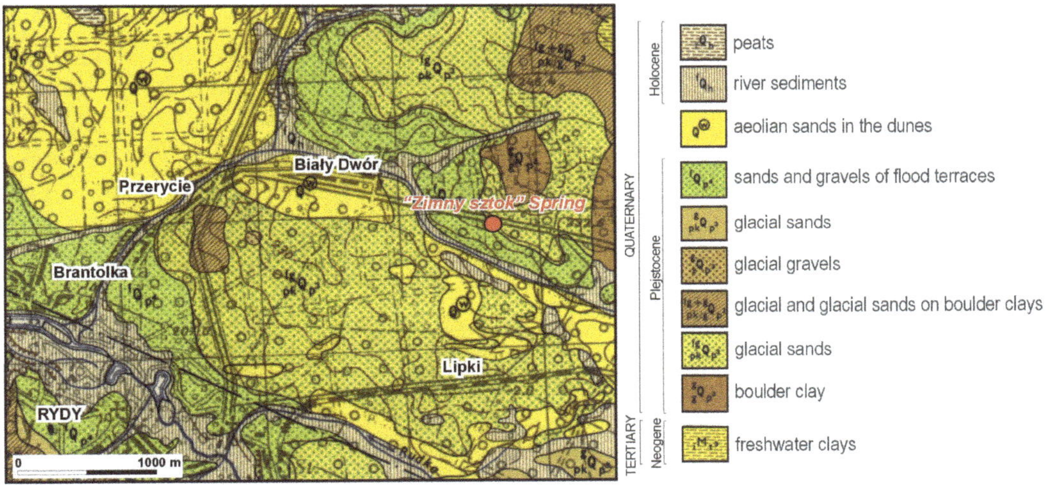

Figure 2. Geological map of the study area [37].

According to the hydrogeological map, the spring is located in an area with a very low risk to water quality. There are two aquifers in the research area in question—the Quaternary and the Neogene. The Quaternary is associated with layers of sand and gravel of the river or glacial formations, as well as intra-moraine sands of various extents and thicknesses. The upper level is located in the upper part of the Quaternary sediments; it often has the character of suspended water at various depths and is underlain by semi-permeable Quaternary clay or impermeable Neogene clay. The second level is found at a depth of approximately 7–10 m and is located on impermeable glacial, Quaternary clay or directly on Miocene clay. Isolating layers, unlike aquifers, are more often spatially continuous. The Quaternary water supply is supplied directly from atmospheric precipitation, and the basis of drainage here is surface streams. The usable Neogene aquifer is associated with sand and sand–gravel inserts, 7–22 m thick, located in the Sarmatian clay complex. Water utility resources in the Neogene zone have been identified to a depth of approximately 90–100 m, and these are sub-artesian waters. The aquifers are recharged directly from the ground surface outcrops of permeable Sarmatian formations and through permeable Quaternary formations. The flow of water in these aquifers is towards the west [37] (Figure 3).

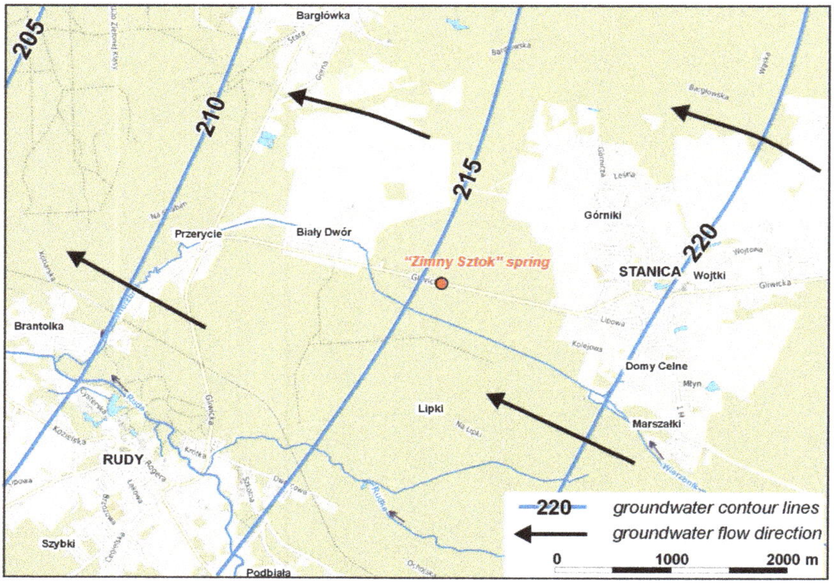

Figure 3. Contour map of the study area [37].

Quaternary waters are characterized by high values of total mineralization, reaching up to 1200 mg/L. They belong to the sulfate–bicarbonate–magnesium–calcium type. Due to the increased concentrations of sulfates, iron, manganese and nitrates, they are classified as water quality class III.

The waters found in the Tertiary sand and gravel complex are fresh waters with mineralization up to 700 mg/L. These waters are of the bicarbonate–calcium and bicarbonate–calcium–magnesium types. Due to exceeding the permissible values of iron and manganese, they are classified as quality class III.

3. Methodology

Research in the Zimny Sztok spring included two measurement series conducted in November 2023 (autumn) and February 2024 (winter). During field research, water samples were collected for physicochemical and bacteriological analysis, which was performed in an accredited laboratory. In terms of physicochemical parameters, the values of electrical conductivity, pH, Ca, Na, K, Mg, Fe, Al, Mn, Ni, Cu, Sr, S, Cl, SO_4, HCO_3, NO_3, NO_2, NH_4, PO_4, N, TOC, Pb, Cd, Cr, Hg, Zn, acidity and alkalinity were obtained. The potential presence of pathogens in environmental waters was inferred based on the determination of coliforms, enterococci, *P. aeruginosa* and *C. perfringens*. The results were compared to values from the Regulation of the Minister of Health in 7 December 2017 on the quality of water intended for human consumption, which are publicly available on the Internet in the Journal of Laws 2017, item 2294.

The Water Quality Index was calculated for both series in order to determine the risk to water quality. The WQI was calculated in relation to values stated in the Guidelines for drinking-water quality (GDWQ) proposed by the WHO. This index was used as the one of the most representative indexes for water risk assessment [38]. Calculating this index includes the following stages: (i) assigning weights to physicochemical parameters, (ii) developing a rating scale and (iii) calculating the WQI.

The WQI was calculated for the six parameters and assigned appropriate relative weights on the basis of the literature [38], dissolved oxygen (0.22), total number of microorganisms at 22 ± 2 °C (0.21), pH (0.15), total phosphate (0.13), nitrates (0.13) and EC (0.16).

Following this, the quality rating scale (q_i) was calculated using Equation (1).

$$q_i = \frac{C_i}{S_i} \cdot 100 \quad (1)$$

where

C_i is the concentration of parameters;

S_i is the standard value of parameters.

The WQI is calculated using the following equation:

$$WQI = \sum_{n=1}^{N} SI \quad (2)$$

where

SI is the subindex calculated from Equation (3).

$$SI = W_i \cdot q_i \quad (3)$$

The index classifies water into the following classes: WQI < 50, excellent water; 50–100 WQI, good water; and WQI > 100, poor water.

4. Results and Discussion

The division into the supply of another hydrographic facility is surface discharge. A gushet is a spring sphere of discharge. The spring yield was measured in both measurement series. The yield of spring was equal to 94 L/h in November 2023 and 99 L/h in February 2024.

Taking into account the content of main ions, the chemical type of water from both measurement series was determined. In the first series of measurements, the water from this spring is classified as bicarbonate–calcium, and in the second series, it is classified as bicarbonate–calcium–magnesium water.

The resulting biological and physicochemical values were compared to the national quality standards of water intended for human consumption (Regulation of the Minister of Health of 7 December 2017 on the quality of water intended for human consumption), limits for the first class of water quality (Regulation of the Minister of Maritime Economy and Inland Navigation of 11 October 2019 on the criteria and method of assessing the status of groundwater bodies) and to the Guidelines for drinking-water quality (GDWQ) proposed by the WHO.

Overall, taking into account the results of both chemical analyses and relating them to the national standards for drinking water, it was revealed that the standards were not met for pH, dissolved oxygen, total number of microorganisms at 22 ± 2 °C and total number of microorganisms at 36 ± 2 °C. When we compare the results to the first-quality class, the standards are not met for pH, nitrates and sulfates. Additionally, copper is at the limit. When comparing test values with the WHO's standards, the situation is identical to national standards for drinking water, except for the dissolved oxygen value.

The value of specific electrolytic conductivity was 208 µS/cm in the first measurement series and 217 µS/cm in the second series. Electrical conductivity (EC) is a parameter used to indirectly assess water mineralization, which, in shallow waters exposed to anthropogenic pollution, may exceed the value of 1000 µS/cm [39]. These waters were characterized by a reduced pH value, which is typical for this area. Unfortunately, excessive consumption of acidified water may cause health consequences such as gout, hypertension, inflammation of the digestive system's organs and skin diseases. However, a lower pH can inhibit the growth of bacteria such as E. coli, due to the presence of chlorine [40]. Regarding this parameter, the pH did not reach the appropriate value, but water quality tests carried out for this spring will continue in the spring and summer. The DO values in the first and second measurement series were lower than the standards. In the first series, in which <300 in the total number of microorganisms was at 22 ± 2 °C, the DO value was lower, suggesting that the bacteria consumed surrounding oxygen [41].

Water pollution with nitrates is primarily caused by their use in agricultural fertilizers to increase crop yields. Nitrogen fertilizers are the main source of water-soluble nitrates and nitrite compounds in the soil [42]. Additionally, increased nitrate content in water may be the result of contamination by human, animal and industrial waste [43]. It should be noted that water in wells, shallow groundwater and water in springs are subject to increased nitrate contamination [44]. Despite the fact that the nitrate content in the Zimny Sztok spring did not exceed the permissible limit for drinking water, the content classifying these waters as the first quality class was exceeded. Due to the fact that the spring is located next to the road, the amount of nitrates should also be controlled. An increased supply of nitrates can be carcinogenic [45]. The content of nitrates in the water should also be controlled due to the proximity of single-family houses that are not fully connected to the sewage system and because nitrates may migrate to the water from leaking septic tanks.

As mentioned before, taking into account the permissible water quality limits for the first class (Tables 1 and 2), copper is exactly on the border between the first and second classes. There is no value for copper in the national standards for drinking water, and the WHO defines the standard as 2 mg/L. Despite this fact, it should be taken into account that this indicator would reduce the quality of water from this source. Excessive concentration of this ion may also cause heavy metal toxicity, which may result in hypertension, kidney failure, liver failure and brain dysfunction [46].

Current national legal acts require that drinking water be analyzed in terms of the total number of microorganisms present in it. The total number of microorganisms growing at 36 ± 2 °C may suggest the presence of pathogenic bacteria (those growing at human body temperature, mesophilic bacteria), while the total number of microorganisms growing at 22 ± 2 °C (psychrophilic bacteria), similarly to the number of coliform bacteria, is used to assess the condition of the water supply network and the operation of treatment systems [47]. Legal solutions refrain from specifying a categorical value of this parameter, requiring its monitoring in order to detect results that deviate from the normal state, which is estimated based on repeated tests. Therefore, it is important and appropriate to conduct further microbiological research in the waters of this spring. It should be added that, during laboratory analysis, the presence of the sought indicator microorganisms in a water sample depends on many factors. These include, among others, the condition of the water supply network, collection technique, storage conditions, analytical method and the human factor. Periodic detection of the presence of indicator bacteria in drinking water may be related to the contamination of water intake with feces and insufficient treatment and disinfection, as well as other reasons—for example, the multiplication of microorganisms inside the water supply network in the form of the so-called biofilm. In the event of increased water flow, the biofilm may detach from the spring outlet and penetrate into the analyzed sample. In the second measurement series (lower temperatures of water and air), no bacteria were observed in the water. It should also be noted that total number of microorganisms at 36 ± 2 °C were also observed in the first sample. Their content was over three times higher than the permissible standards. Because this test suggests that the water contains bacteria that prefer to grow at body temperature, drinking such water may be dangerous; however, it must be confirmed that this level of bacteria in this water persists.

Table 1. Results of the chemical analysis performed.

Physicochemical Parameters	Unit	November 2023	February 2024	Limits for Drinking Water [1]	Limits for the First Class of Quality [2]	The WHO's Standards [3]	Limit of Quantification
EC	µS/cm	208	217	2500	700	400	<10–12,000>
Na	mg/L	7.1	9	200	60		<1–100,000>
K	mg/L	6.5	5.9	-	10		<1–10,000>
Ca	mg/L	24	32	-	50		<0.1–10,000>
Mg	mg/L	4.1	6.8	125	30		<0.1–5000>
Fe	mg/L	0.002	0.073	0.2	0.2		<0.01–500>
Al	mg/L	0.005	0.005	0.2	0.1		<0.05–50>
Mn	mg/L	0.007	0.014	0.05	0.05	0.4	<0.005–100>
Ni	mg/L	0.002	0.002	0.002	0.005	0.07	<0.01
Cu	mg/L	0.002	0.01	-	0.01	2	<0.005–100>
Sr	mg/L	0.11	0.13	-			<0.02–500>
S	mg/L	15	17	-			<0.1–150>
Cl	mg/L	18	18	250	60	200	<1–10,000>
SO_4	mg/L	57	61	250	60		<1–10,000>
HCO_3	mg/L	19.5	33.3	-	200		<6.1–6100>
NO_3	mg/L	13	11	50	10	50	<0.44–443>
NO_2	mg/L	0.033	0.033	0.5	0.03	3	<0.033–33>
NH_4	mg/L	0.065	0.065	0.5	0.5		<0.04–2576>
PO_4	mg/L	0.07	0.062	-	0.5		<0.15–100>
N_K	mg/L	2.5	2.5	-		1.5	<1–1000>
TOC	mg/L	1	1	-	5		<1–1000>
Pb	mg/L	0.002	0.002	0.01	0.01	0.01	<0.01–100>
Cd	mg/L	0.00025	0.002	0.005	0.001	0.003	<0.0002–1>
Cr	mg/L	0.0015	0.00025	0.05	0.01	0.05	
Hg	mg/L	0.00005	0.0015	0.001	0.001	0.006	<0.0001–0.01>
Zn	mg/L	0.0025	0.019	-	0.05		<0.02–500>
Acidity	mmol/L	0.1	4.3	-			<0.1–100>
Alkalinity	mg/L $CaCO_3$	16	27.3	-			<5–5000>
Dissolved oxygen (DO)	mg/L	4.1	5.6	8			-
pH		5.54	6.05	6.5–8.5	6.5–9.5	6.5–8.5	<2–12>

[1] Regulation of the Minister of Health of 7 December 2017 on the quality of water intended for human consumption. [2] (Regulation of the Minister of Maritime Economy and Inland Navigation of 11 October 2019 on the criteria and method of assessing the status of groundwater bodies. [3] The Guidelines for drinking-water quality (GDWQ) proposed by the WHO [41].

Table 2. Results of the microbiological analysis performed.

Microbiological Parameters	Unit	November 2023	February 2024	Limits for Drinking Water [1]	Limits for the First Class of Quality [2]	The WHO's Standards [3]	Limit of Quantification
The number of coliform bacteria	cfu/100 mL	0	0	0		0	≥0
Number of Enterococci	cfu/100 mL	0	0	0		0	≥0
Number of Clostridium perfringens	cfu/100 mL	0	0	0		0	≥0
Total number of microorganisms at 22 ± 2 °C	cfu/mL	>300	0	100		100	≥0
Number of Pseudomonas aeruginosa	cfu/100 mL	0	0	0		0	≥0
Total number of microorganisms at 36 ± 2 °C	cfu/mL	68	0	20		20	≥0

[1] Regulation of the Minister of Health of 7 December 2017 on the quality of water intended for human consumption. [2] (Regulation of the Minister of Maritime Economy and Inland Navigation of 11 October 2019 on the criteria and method of assessing the status of groundwater bodies. [3] The Guidelines for drinking-water quality (GDWQ) proposed by the WHO [41].

In addition to comparing the obtained analytical results to all mentioned standards, WQI values were calculated for both series. When calculating the WQI, data on dissolved oxygen, the total number of microorganisms at a temperature of 22 ± 2 °C, pH, total phosphates, nitrates and EC were used. The calculated values of the WQI show that one parameter can significantly influence its total value (Figures 4 and 5). In the first series of measurements, in which it was found that the total number of microorganisms at a temperature of 22 ± 2 °C was greater than 300 CFU/mL, with the acceptable standard being 100, the total WQI value was over 60% higher than if this parameter had not been taken into account. The total value of the indicator in the first measurement series was 99.67. The WQI result (close to 100) obtained for the Zimny Sztok source from the first measurement series can be compared with WQI values from the area of typical waste landfills [48]. In the case of drinking water sources, the values of this indicator should be in the range of 0–25 [49,50].

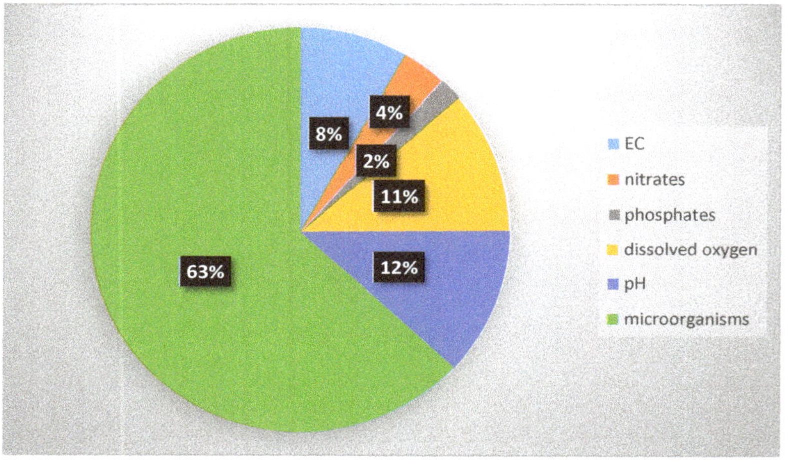

Figure 4. The share of individual parameters in the WQI values in the first measurement series.

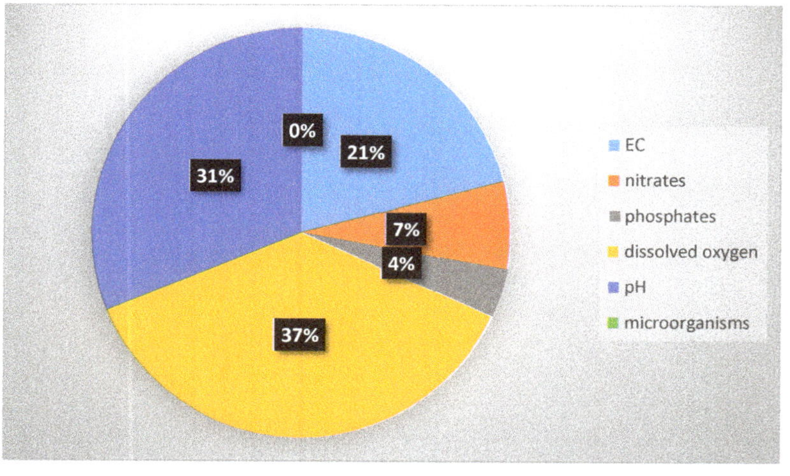

Figure 5. The share of individual parameters in the WQI values in the second measurement series.

Such a high WQI value indicates a high level of water pollution. The calculated WQI value for the second series of measurements indicates that the water in the spring is of excellent quality (WQI is less than 50; Figure 5). Such a large change in the index value is caused by the lack of presence of microorganisms. In the second series of measurements, the influence of EC and pH is more visible. It can be concluded that for water risk assessment, the result from the second series is more reliable. Analyzing the calculated WQI values for both measurement series, it should be noted that the content of dissolved oxygen and the pH value have a large share in the final value of the indicator.

The results of the WQI calculations indicate that better water quality was recorded in the second measurement series. Meanwhile, the very high value of the indicator in the first measurement series results, as already mentioned, from the large total number of microorganisms at a temperature of 22 ± 2 °C. If the WQI value was calculated by changing this parameter to another physicochemical parameter, it would turn out that the quality of this water is better, compared to the second measurement series. Having the obtained results of physicochemical and bacteriological analyzes at our disposal, the data for this bacteriological parameter should be omitted in the calculation of the WQI. It should be noted that the occurrence of bacteria in the waters of springs such as Zimny Sztok is often dependent on the episodic inflow of pollutants from leaky septic tanks, animal activity or other anthropogenic activities. The first step before further water measurements in this source and in the use of water by residents should be to protect this source against the inflow of pollutants from the surface. Based on the analysis of the spatial development map of the immediate area, it can be concluded that the main sources of pollution may be fertilizers from agricultural fields, leaky septic tanks and pollutants migrating from the nearby road.

Analyzing time changes for individual parameters, the largest increases were observed in the case of mercury, zinc, cadmium, iron, bicarbonates and sodium. In the case of mercury between November and February, the content increased 30 times. Despite the fact that mercury concentrations in the tested source are very low, such an increase should be taken into account in any subsequent measurement because heavy metals naturally occurring in the aquatic environment in higher concentrations may be toxic when they accumulate in organisms [51]. Changes in water pH are related to the metal content. Even though the pH value in water was lower in the first measurement series, no higher concentrations of zinc or copper were observed in this series. As the Zimny Sztok spring is located in an urban space, higher concentrations of metals in water in the second measurement series can be correlated with increased heating of apartments and the use of various heating sources, from which pollutants may migrate to the atmosphere and from there to water. In terms of hardness, the water in the spring is classified as soft. This condition persists even after the alkalinity value increased by 70% in the second measurement series. A decrease in pollutants in the second measurement series was noticeable only in the case of chromium (a decrease of approximately 83%), nitrates (a decrease of approximately 18%), phosphates (a decrease of approximately 13%) and potassium (a decrease of approximately 9%).

In terms of factors such as season, sampling methods, geology conditions and anthropogenic activities, it should be noted that only the air temperature and anthropogenic activities related to heating apartments, limiting the amount of fertilizers used, and changing the number of inhabitants using the source could have an influence on the value of the WQI. The method of collecting samples for analysis was identical in both cases, and the analysis was performed in the same accredited laboratory.

It is worth remembering that, in the world, about 65% of groundwater is used for drinking and irrigation, 20% of this water is used for feeding animals and about 15% of groundwater is used for industry purposes [13]. Poland is one of the countries that primarily uses groundwater resources for drinking purposes. Springs are a valuable complement to these resources; therefore, it is necessary to use them rationally. The described spring has existed since the 19th century. Due to the fact that the Zimny Sztok spring is located by the road and is protected by a concrete casing, it is necessary to take a number of actions to ensure that these waters are sustainably used. It should be mentioned that there is no continuous monitoring of water quality at this source. However, this point is not included in the observation network of the State Environmental Monitoring program, which would certainly be a guarantee of obtaining more quality data. Unfortunately, monitoring studies in springs are rarely performed, which is due to the fact that piezometers and wells dominate the system. In order to use resources sustainably, it would also be necessary to secure the area around the source and the intake point itself, control leaks in septic tanks in the commune, limit the possibility of the migration of pollutants from fertilizers and also undertake educational activities for local residents and other users in order to present the current chemical and bacteriological composition of these waters, determining the consequences of consuming water that does not meet the appropriate standards for drinking water and counteracting anthropogenic threats.

5. Conclusions

The aim of this study was achieved by conducting tests in spring waters, presenting the results of these tests, calculating the WQI and indicating what further course of action should be taken in the field of monitoring these waters.

Based on the tests of physicochemical parameters, it can be concluded that the conductivity and most of the parameters in water samples from the Zimny Sztok spring are within the safe limits according to the WHO's guidelines and are also at a safe drinking level based on national standards. The only exceedances concern nitrates and copper ions. The pH of the water is lowered by approximately 1 point compared to the lower limit of standards, which is typical for the entire region.

The obtained results indicate a certain difference in concentrations in both measurement series, e.g., a 30-fold increase in the mercury concentration in the second measurement series or an 8-fold increase in cadmium concentration. In terms of microbiological parameters, in the first measurement series, >300 cfu/mL was recorded for the total number of microorganisms at a temperature of 22 ± 2 °C and 68 cfu/mL for the total number of microorganisms at a temperature of 36 ± 2 °C. Due to the fact that during laboratory analysis, the presence of the sought indicator microorganisms in a water sample depends on many factors, research on these two parameters should be continued to confirm the origin of these values.

WQI values were different in both measurement series. In the first series of measurements, the WQI was over 99, and in the second series, it was approximately 41. Importantly, such a high value of the indicator in the first series is the result of the high value of the total number of microorganisms at a temperature of 22 ± 2 °C. Even though the WQI is one of the most optimal indicators for assessing water quality and risk, in the analyzed case, its value might have caused an incorrect analysis of water quality changes in November and February. Further periodic or permanent tests of water quality in terms of physicochemical and bacteriological parameters, including spring parameters, should be conducted by the commune's authorities to ensure the reliability and representativeness of the test results. Additionally, thanks to measurements performed in different seasons of the year, it will be possible to determine the seasonality of changes. Conducting geochemical modeling to determine speciation and saturation rates will also be a valuable element of further research.

The relationship between the urban space and water determines the healthy prosperity and sustainable development of such areas. In this context, it is important to conduct rational management of water resources but also formulate development strategies based on the state of the water resources in question. In order to maintain appropriate quantitative and qualitative resources of water intended for consumption in urban spaces, it is necessary to monitor and control parameters, as well as analyze the possible effects of consuming such water. Such actions should also be taken in the case of the Zimny Sztok spring. This is important not only in the case of water that is introduced into the system but also, above all, in the case of water that is collected by local residents from springs.

Author Contributions: Conceptualization, M.R. and D.D.; methodology, M.R.; software, D.D.; investigation, D.D. and M.R.; data curation, M.R. and D.D.; writing—original draft preparation, D.D. and M.R. All authors have read and agreed to the published version of the manuscript.

Funding: This research received no external funding.

Institutional Review Board Statement: Not applicable.

Informed Consent Statement: Not applicable.

Data Availability Statement: There are no other data.

Conflicts of Interest: The authors declare no conflicts of interests.

References

1. Barton, H.; Grant, M. A Health Map for the Local Human Habitat. *J. R. Soc. Promot. Health* **2006**, *126*, 252–253. [CrossRef] [PubMed]
2. Ki, D.; Lee, S. Analyzing the Effects of Green View Index of Neighborhood Streets on Walking Time Using Google Street View and Deep Learning. *Landsc. Urban Plan.* **2021**, *205*, 103920. [CrossRef]
3. Tonne, C.; Adair, L.; Adlakha, D.; Anguelovski, I.; Belesova, K.; Berger, M.; Brelsford, C.; Dadvand, P.; Dimitrova, A.; Giles-Corti, B.; et al. Defining Pathways to Healthy Sustainable Urban Development. *Environ. Int.* **2021**, *146*, 106236. [CrossRef] [PubMed]
4. Sumdang, N.; Chotpantarat, S.; Cho, K.H.; Thanh, N.N. The Risk Assessment of Arsenic Contamination in the Urbanized Coastal Aquifer of Rayong Groundwater Basin, Thailand Using the Machine Learning Approach. *Ecotoxicol. Environ. Saf.* **2023**, *253*, 114665. [CrossRef] [PubMed]
5. Arora, N.K.; Mishra, I. Sustainable Development Goal 6: Global Water Security. *Environ. Sustain.* **2022**, *5*, 271–275. [CrossRef]
6. Arvizu, I.S.; Murray, S.R. A Simple, Quantitative Assay for the Detection of Viable but Non-Culturable (VBNC) Bacteria. *STAR Protoc.* **2021**, *2*, 100738. [CrossRef] [PubMed]
7. Thapa, B.; Pant, R.R.; Thakuri, S.; Pond, G. Assessment of spring water quality in Jhimruk RiverWatershed, Lesser Himalaya, Nepal. *Environ. Earth Sci.* **2020**, *79*, 504. [CrossRef]
8. Hernandez-Pastor, L.; Geurtsen, J.; Baugh, B.; El Khoury, A.C.; Kalu, N.; Krishnarajah, G.; Gauthier-Loiselle, M.; Bungay, R.; Cloutier, M.; Saade, E. Economic Burden of Invasive *Escherichia coli* Disease among Older Adult Patients Treated in Hospitals in the United States. *J. Manag. Care Spec. Pharm.* **2023**, *29*, 873–883. [CrossRef] [PubMed]
9. Wang, L.; Zhang, L.; Lv, J.; Zhang, Y.; Ye, B. Public Awareness of Drinking Water Safety and Contamination Accidents: A Case Study in Hainan Province, China. *Water* **2018**, *10*, 446. [CrossRef]
10. Backes, J.G.; Traverso, M. Life Cycle Sustainability Assessment as a Metrics towards SDGs Agenda 2030. *Curr. Opin. Green Sustain. Chem.* **2022**, *38*, 100683. [CrossRef]
11. Hang, L.T.T.; Huong, L.T.T. Domestic Use of Rainwater in Ho Chi Minh City, Vietnam: Exploring the Barriers from the Citizens' Perspective. *J. Water Clim. Chang.* **2019**, *10*, 210–222. [CrossRef]
12. Owamah, I.H.; Asiagwu, A.K.; Egboh, S.H.O.; Phil-Usiayo, S. Drinking Water Quality at Isoko North Communities of the Niger Delta Region, Nigeria. *Toxicol. Environ. Chem.* **2013**, *95*, 1116–1128. [CrossRef]
13. Li, Y.; Cai, Y.; Wang, X.; Li, C.; Liu, Q.; Sun, L.; Fu, Q. Classification Analysis of Blue and Green Water Quantities for a Large-Scale Watershed of Southwest China. *J. Environ. Manag.* **2022**, *321*, 115894. [CrossRef] [PubMed]
14. Falkenmark, M.; Rockström, J. The New Blue and Green Water Paradigm: Breaking New Ground for Water Resources Planning and Management. *J. Water Resour. Plan. Manag.* **2006**, *132*, 129–132. [CrossRef]
15. Yan, W.; Li, F.; Zhao, Y. Integrating Reasonable Proportion of Blue and Green Water into Land Use Planning—A New Ecological Perspective. *Ecohydrology* **2022**, *15*, e2459. [CrossRef]
16. Liu, S.; Wang, F.; Wang, X.; Luo, H.; Wang, L.; Zhou, P.; Xu, C. Modeling the Impact of Climate Change on Streamflow in Glacier/Snow-Fed Northern Tianshan Basin. *J. Hydrol. Reg. Stud.* **2023**, *50*, 101552. [CrossRef]
17. Zhu, L.; Gong, H.; Li, X.; Wang, R.; Chen, B.; Dai, Z.; Teatini, P. Land Subsidence Due to Groundwater Withdrawal in the Northern Beijing Plain, China. *Eng. Geol.* **2015**, *193*, 243–255. [CrossRef]
18. Khatri, N.; Tyagi, S. Influences of Natural and Anthropogenic Factors on Surface and Groundwater Quality in Rural and Urban Areas. *Front. Life Sci.* **2015**, *8*, 23–39. [CrossRef]
19. Chen, Z.; Auler, A.S.; Bakalowicz, M.; Drew, D.; Griger, F.; Hartmann, J.; Jiang, G.; Moosdorf, N.; Richts, A.; Stevanovic, Z.; et al. The World Karst Aquifer Mapping Project: Concept, Mapping Procedure and Map of Europe. *Hydrogeol. J.* **2017**, *25*, 771–785. [CrossRef]
20. Twinomucunguzi, F.R.B.; Nyenje, P.M.; Kulabako, R.N.; Semiyaga, S.; Foppen, J.W.; Kansiime, F. Emerging Organic Contaminants in Shallow Groundwater Underlying Two Contrasting Peri-Urban Areas in Uganda. *Environ. Monit. Assess.* **2021**, *193*, 228. [CrossRef] [PubMed]
21. Bamsaoud, S.F.; Bin Saeed, F.F. Physical and Chemical Characteristics of Assaiq and Senah Hot Springs Water in Hadhramout-Yemen and the Assessment of Water Quality for Drinking and Irrigation Purposes. *J. Phys. Conf. Ser.* **2021**, *1900*, 012011. [CrossRef]

22. Todd, D.K. *Groundwater Hydrology: International Edition*, 3rd ed.; John Wiley and Sons (WIE): Brisbane, QLD, Australia, 2004.
23. Keeler, B.L.; Polasky, S.; Brauman, K.A.; Johnson, K.A.; Finlay, J.C.; O'Neill, A.; Kovacs, K.; Dalzell, B. Linking Water Quality and Well-Being for Improved Assessment and Valuation of Ecosystem Services. *Proc. Natl. Acad. Sci. USA* **2012**, *109*, 18619–18624. [CrossRef] [PubMed]
24. Gothwal, R.; Shashidhar, T. Antibiotic Pollution in the Environment: A Review. *Clean* **2015**, *43*, 479–489. [CrossRef]
25. Grimalt, J.O.; Fernandez, P.; Berdie, L.; Vilanova, R.M.; Catalan, J.; Psenner, R.; Hofer, R.; Appleby, P.G.; Rosseland, B.O.; Lien, L.; et al. Selective Trapping of Organochlorine Compounds in Mountain Lakes of Temperate Areas. *Environ. Sci. Technol.* **2001**, *35*, 2690–2697. [CrossRef] [PubMed]
26. Kumar, R.; Sharma, P.; Manna, C.; Jain, M. Abundance, Interaction, Ingestion, Ecological Concerns, and Mitigation Policies of Microplastic Pollution in Riverine Ecosystem: A Review. *Sci. Total Environ.* **2021**, *782*, 146695. [CrossRef]
27. Chrysanthus, N. Bacteriological Quality of Alternative Water Sources in Bambui and Bambili Residential Areas, North-West Region, Cameroon. *OAlib* **2014**, *1*, 1–6. [CrossRef]
28. Chartres, N.; Bero, L.A.; Norris, S.L. A Review of Methods Used for Hazard Identification and Risk Assessment of Environmental Hazards. *Environ. Int.* **2019**, *123*, 231–239. [CrossRef] [PubMed]
29. Frank, S.; Fahrmeier, N.; Goeppert, N.; Goldscheider, N. High-Resolution Multi-Parameter Monitoring of Microbial Water Quality and Particles at Two Alpine Karst Springs as a Basis for an Early-Warning System. *Hydrogeol. J.* **2022**, *30*, 2285–2298. [CrossRef]
30. Mukherjee, S.; Ashfaq, M.; Mishra, A.K. Compound Drought and Heatwaves at a Global Scale: The Role of Natural Climate Variability-associated Synoptic Patterns and Land-surface Energy Budget Anomalies. *J. Geophys. Res.* **2020**, *125*, e2019JD031943. [CrossRef]
31. Hinkel, J. "Indicators of Vulnerability and Adaptive Capacity": Towards a Clarification of the Science–Policy Interface. *Glob. Environ. Chang.* **2011**, *21*, 198–208. [CrossRef]
32. Oke, S.A.; Vermeulen, D.; Gomo, M. Aquifer Vulnerability Assessment of the Dahomey Basin Using the RTt Method. *Environ. Earth Sci.* **2016**, *75*, 964. [CrossRef]
33. Juntunen, J.; Meriläinen, P.; Simola, A. Public Health and Economic Risk Assessment of Waterborne Contaminants and Pathogens in Finland. *Sci. Total Environ.* **2017**, *599–600*, 873–882. [CrossRef] [PubMed]
34. Szkleniarz, J. Application of Selected Groundwater Quality Indices in Anthropogenically Transformed Areas. *Prace KGP PTG* **2024**, *38*, 43–61. [CrossRef]
35. Kondracki, J. *Geografia Regionalna Polski*; Wydawnictwo Naukowe PWN: Warszawa, Poland, 2000.
36. Czarnecka, H. *Atlas of the Hydrographic Division of Poland*; IMGW: Suwalki, Poland, 2005.
37. Chmura, A. *Mapa Hydrogeologiczna Polski w Skali 1:50,000 Arkusz Gliwice*; Państwowy Instytut Geologiczny: Warszawa, Poland, 1998.
38. Kumar, M.; Ramanathan, A.L.; Tripathi, R.; Farswan, S.; Kumar, D.; Bhattacharya, P. A Study of Trace Element Contamination Using Multivariate Statistical Techniques and Health Risk Assessment in Groundwater of Chhaprola Industrial Area, Gautam Buddha Nagar, Uttar Pradesh, India. *Chemosphere* **2017**, *166*, 135–145. [CrossRef] [PubMed]
39. Vitale, R.J.; Braids, O.C. *Practical Handbook of Environmental Site Characterization and Ground-Water Monitoring*, 2nd ed.; Nielsen, D.M., Ed.; CRC Press: Boca Raton, FL, USA, 2006.
40. Deng, M. Theoretical Analysis of "Natural-Social-Trading" Ternary Water Cycle Mode in the Inland River Basin of Northwest China. *Acta Geograph. Sin* **2020**, *56*, 1333–1345.
41. Harichandan, A.; Sekhar Patra, H.; Mohan Sethy, K. Evaluation of Water Quality of Local Streams at Gandhamardan Iron Mines, Suakati, Keonjhar District of Odisha, India. *J. Pollut. Eff. Control* **2017**, *5*, 199. [CrossRef]
42. Beeckman, F.; Motte, H.; Beeckman, T. Nitrification in Agricultural Soils: Impact, Actors and Mitigation. *Curr. Opin. Biotechnol.* **2018**, *50*, 166–173. [CrossRef] [PubMed]
43. Shukla, S.; Saxena, A. Sources and Leaching of Nitrate Contamination in Groundwater. *Curr. Sci.* **2020**, *118*, 883. [CrossRef]
44. Lundberg, J.O.; Carlström, M.; Weitzberg, E. Metabolic Effects of Dietary Nitrate in Health and Disease. *Cell Metab.* **2018**, *28*, 9–22. [CrossRef] [PubMed]
45. Ward, M.H.; Jones, R.R.; Brender, J.D.; de Kok, T.M.; Weyer, P.J.; Nolan, B.T.; Villanueva, C.M.; van Breda, S.G. Drinking Water Nitrate and Human Health: An Updated Review. *Int. J. Environ. Res. Public Health* **2018**, *15*, 1557. [CrossRef] [PubMed]
46. Wang, Z.; Kong, D.; Qiao, N.; Wang, N.; Wang, Q.; Liu, H.; Zhou, Z.; Ren, Z. Facile Preparation of Novel Layer-by-Layer Surface Ion-Imprinted Composite Membrane for Separation of Cu^{2+} from Aqueous Solution. *Appl. Surf. Sci.* **2018**, *457*, 981–990. [CrossRef]
47. Admin. "Woda Ze Studni—Rodzaje, Przyczyny I Skutki Zanieczyszczeń." Inżynieria Wody, 21 March 2024. Available online: https://inzynieriawody.pl/woda-ze-studni-rodzaje-przyczyny-i-skutki-zanieczyszczen/ (accessed on 14 May 2024).
48. Ferreira, C.S.S.; Adama-Ajonye, O.; Ikenna, A.E.; Kalantari, Z. Groundwater Quality in the Vicinity of a Dumpsite in Lagos Metropolis, Nigeria. *Geogr. Sustain.* **2023**, *4*, 379–390. [CrossRef]
49. Taloor, A.K.; Pir, R.A.; Adimalla, N.; Ali, S.; Manhas, D.S.; Roy, S.; Singh, A.K. Spring Water Quality and Discharge Assessment in the Basantar Watershed of Jammu Himalaya Using Geographic Information System (GIS) and Water Quality Index(WQI). *Groundw. Sustain. Dev.* **2020**, *10*, 100364. [CrossRef]

50. Pantha, S.; Timilsina, S.; Pantha, S.; Manjan, S.K.; Maharjan, M. Water Quality Index of Springs in Mid-Hill of Nepal. *Environ. Chall.* **2022**, *9*, 100658. [CrossRef]
51. Kumar, M.; Singh, S.; Dwivedi, S.; Trivedi, A.; Dubey, I.; Trivedi, S.P. Copper-Induced Genotoxicity, Oxidative Stress, and Alteration in Transcriptional Level of Autophagy-Associated Genes in Snakehead Fish Channa Punctatus. *Biol. Trace Elem. Res.* **2023**, *201*, 2022–2035. [CrossRef] [PubMed]

sustainability

MDPI

Article

Effect of Soil Erosion on Soil and Plant Properties with a Consequence on Related Ecosystem Services

Radoslava Kanianska [1], Miriam Kizeková [2,*], Ľubica Jančová [2], Jozef Čunderlík [2] and Zuzana Dugátová [2]

[1] Faculty of Natural Sciences, Matej Bel University in Banská Bystrica, Tajovského 40, 974 01 Banská Bystrica, Slovakia; radoslava.kanianska@umb.sk

[2] Grassland and Mountain Agriculture Institute, National Agricultural and Food Centre, Mládežnícka 36, 974 21 Banská Bystrica, Slovakia; lubica.jancova@nppc.sk (Ľ.J.); jozef.cunderlik@nppc.sk (J.Č.); zuzana.dugatova@nppc.sk (Z.D.)

* Correspondence: miriam.kizekova@nppc.sk

Abstract: Erosion is a process often driven by land management deteriorating or changing soil properties along the slopes, with consequences on ecosystem services. In a model area with Stagnic Cambisol, with two different types of land use (grassland—GL and arable land—AL), on an erosion transect in three different hillslope positions (upper, middle, and lower), in two different depths (0–10 and 35–45 cm), we observed the impact of soil erosion on soil and plant properties and ecosystem services by use of direct measurements and models. In GL, soil available potassium (SK), soil available phosphorus (SP) and pH increased both downward along the slope and in soil depth. A significantly ($p < 0.01$) higher content of plant nutrients (PN, PP, and PK) and shoot biomass was recorded in the lower part of the hillslope. In AL, soil parameters (pH, SOC, SN, and SOC/SN) reached the lowest values at the middle hillslope position at the shallowest depth. A relatively negligible annual average soil loss was recorded for GL (0.76 t/ha/yr). To the contrary, a very high rate of soil erosion was found for AL with maize silage. The actual soil moisture was 50% higher in GL compared to AL, which was reflected also in the soil water deficit index (SWDI) being more favorable for GL.

Keywords: soil erosion; grassland; arable land; hillslope position; soil ecosystem service

1. Introduction

Soil is a non-renewable and precious natural resource; it is crucial to sustainability and vital to human's existence. Soil is a fundamental part of the Earth's ecosystems [1]. It supports crop and animal production and plays a critical role in delivering many ecosystem services. Nevertheless, soils are jeopardized by a wide range of natural processes or human activities, such as long-lasting rainfalls or various forms of intensive land and agricultural use.

Erosion is a natural phenomenon caused by several factors, such as by the wind and especially by water [2]. It is thanks to erosion that some of the most fertile soils of plains and mountains and hilly landscapes have been formed [3]. On the other hand, erosion is accelerated by human activities such as agricultural activities and becomes unsustainable when soil loss exceeds its rate of formation [4].

Sustainable land use is a major challenge of current land management. Land degradation generated by soil erosion is one of the main issues within the Sustainable Development Goals adopted by the United Nations in 2015 [5]. Currently, soil erosion belongs to the most extended soil degradation process in the world [6]. Mountain structures result in high water erosion rates [7]. Intense rainfall and conventional tillage practices coupled with poor soil structure and steep slopes significantly accelerate soil erosion [8]. Soil water erosion is a complex process during which fertile topsoil is disrupted by water, transported away, and deposited on concave parts of slopes [9]. The result of water erosion process can be creation of colluvial soils on concave parts of slopes. The sediments can display various grain size distribution and diverse contents of soil organic matter, iron, and carbonates. In addition, pedoturbation intensifies a continuous mixture of organic and inorganic material [10].

Erosion causes the leaching of organic matter and nutrients, and the deterioration of water quality and drainage systems. The supply of macronutrients such as phosphorus, calcium, nitrogen, and carbon stocks are modulated by erosion [11]. Even slight erosion negatively affects farming conditions and yield [12] and leads to a loss of ecosystem services [13].

In this study, we are looking for an answer to the question of how soil erosion of differently managed lands can affect basic soil and plant properties with an impact on ecosystem services and their potential. The research was performed at a model area with Stagnic Cambisol, at two study sites (two slopes) with different types of land use (grassland and arable land), in the cadastral area Kečovo, eastern Slovakia, on an erosion transect in three different hillslope positions (upper, middle, and lower), in two different depths (0–10 and 35–45 cm). The main objectives of this study were (1) to evaluate the spatial and vertical differences in the soil's physical and chemical properties in three different hillslope positions (upper, middle, and lower), in two different depths (0–10 and 35–45 cm), and under two different land use management strategies, (2) to evaluate plant properties affected by water erosion, (3) to predict potential soil losses due to water erosion using an empirical model (Universal Soil Loss Equation—USLE) under two different land use management strategies, (4) to estimate the potential of soil ecosystem services (carbon stock, nitrogen stock, and water retention) under the influence of water erosion, and (5) to assess how the intensity of management, soil, and environmental factors affect water erosion's impact on soil and plant properties, reflected in ecosystem services' potential.

2. Materials and Methods

2.1. Sites Description

This study was conducted in Slovakia on two short steep (from 10 to 12%) slopes used as grassland (GL) and arable land (AL). Slovakia is mostly a mountain country (55% of the land territory) located in the western Carpathians. The climate is temperate. A high variability of soil types and soil particles over short distances is typical for some Slovak mountain and lowland regions as well [14]. In Slovakia, approximately 39% of agricultural land is threatened by water erosion [15]. The study area with two study sites was situated in the cadastral area Kečovo, located in the Slovak Karst region (part of the Silica plateau) in the Western Carpathians. The Slovak Karst spreads over more than 800 km^2 and its largest area is located between 500 and 700 m a.s.l. The terrain comprises medium height mountains. The altitude difference between the valley and the plateau surfaces is ~400 m [16]. From a geological perspective, the Slovak Karst is a complex of Mesozoic rocks/Wetterstein limestone [17].

The soil type at the study sites was classified according to the World Reference Base for Soil Resources [18] as Stagnic Cambisol (CMst) developed on slope neogenic sediments [19]. The long-term (30-year period) annual average total precipitation is 620 mm and annual average air temperature is 8.6 °C. On the arable land, a monoculture of silage maize was grown. The extensive permanent grassland was cut in June and grazed by suckler cows in the second half of the vegetation season.

At the study sites, strip farming has been the main cultivation practice, adopted since the High Middle Ages until the late 1950s. (Figure 1a). The fields of arable land were mainly oriented across the slope. After collectivization in 50s, the fields were merged into large blocks and thus exposed to the influence of erosion (Figure 1b). At present, study site 1 is used as grassland and study site 2 as arable land. The size of each selected plot was 7200 m^2 (with dimensions of 180 × 40 m). The GPS coordinates are 48°28′33.26″ N, 20°27′57.01″ E of study site no. 1 (center point), and 48°28′26.74″ N, 20°27′37.27″ E of study site no. 2 (center point).

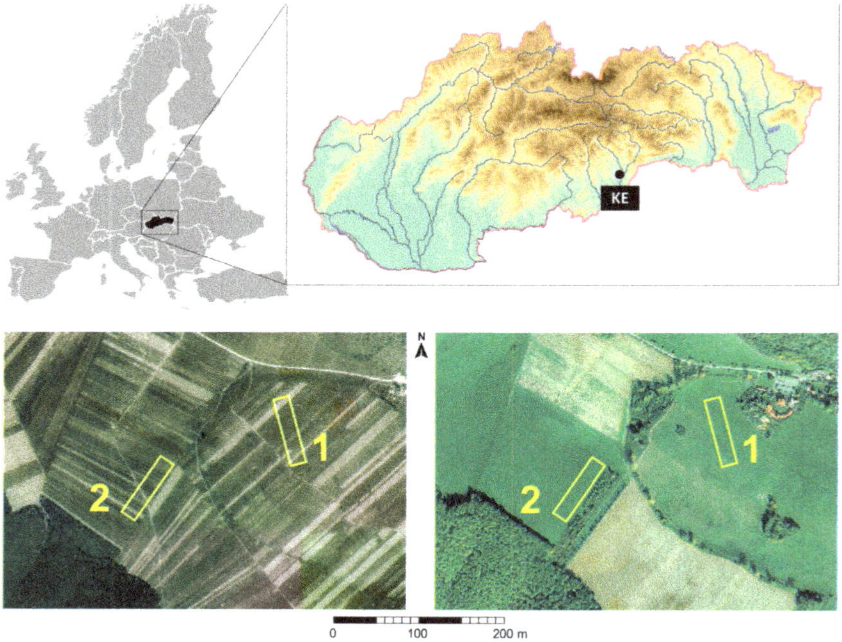

Figure 1. Study area location with two study sites (1, grassland and 2, arable land) and their aerial photographs in (**a**) 1953 (before collectivization) and (**b**) 2021. The historical orthophoto map was created within the project of the Center of Excellence for Support of Decision Making in Forest and Landscape, Technical University in Zvolen and is available at http://mapy.tuzvo.sk (accessed on 30 June 2024). Historical orthophoto map © GEODIS SLOVAKIA, s.r.o. and Historical LMS © Topographic Institute Banská Bystrica. Orthophoto map © EUROSENSE, s.r.o. and GEODIS SLOVAKIA, s.r.o.

2.2. Soil and Plant Sampling and Analysis

We collected soil samples from 0–10 cm and 35–45 cm depths, in order to distinguish soil layers more and less exposed to erosion, from four sampling points with a Z-shaped pattern on an erosion transect in the three different hillslope positions (upper, middle, and lower). The distance between the upper and middle position is 90 m, as is that between the middle and lower positions.

We determined basic soil physical properties in undisturbed soil samples 100 cm^3 in volume that were sampled from both depths using core extracting tubes (Eijkelkamp Equipment for Soil Research, Giesbeek, The Netherlands). The gravimetric method according to Novák [20] was performed to determine bulk density (BD). Particle density (PD) was measured by the pycnometer method according to Blake and Hartage [21]. Physical and hydro-physical properties were calculated by the following formulas:

Bulk density (BD) (Equation (1)),

$$BD = \frac{m_d}{V}\ (\mathrm{g\ cm^{-3}}) \tag{1}$$

where m_d is the mass of dry soil in g and V is the sample volume in the core-extracting tube in cm^3.

Particle density (PD) (Equation (2)),

$$PD = \frac{m_d}{m_w}\ (\text{g cm}^{-3}) \quad (2)$$

where m_d is the mass of dry soil in g and m_w is the mass of water expelled by soil in cm^3.

We conducted the particle size analysis by the pipette method using soil particle sedimentation. Particle size fractions (sand, silt, and clay) were classified according to the United States Department of Agriculture (USDA) system. Penetrometric resistance (PR) was measured directly in the terrain by a penetrologger (Eijkelkamp Equipment for Soil Research, The Netherlands). Soil moisture (SM) and soil temperature (ST) were measured using a WET sensor (Delta-T Devices LTD).

To estimate soil chemical properties, soil samples were air-dried and sifted through a 2 mm mesh sieve. Soil reaction was determined as pH in 1 M KCl solution (with a ratio of 1:2.5). Tjurin method (a modification of Nikitin) [22] was performed to determine soil organic carbon (SOC) content. Soil organic matter (SOM) was recalculated by 1.724 coefficient. Total soil nitrogen (SN) content was determined by the Kjeldahl method [23]. Soil plant-available nutrients K (SK) and P (SP) were extracted by Melich III [24]. SP was measured calorimetrically on the analyzer Scalar and SK by flame photometry. To distinguish soil and plant elements, we used the designation S for soil and P for plant nutrients.

In grasslands, plant samples were taken from the three different hillslope positions at the same sampling points as the soil samples. To assess the shoot dry matter yield, the grassland shoot biomass was clipped in May from twelve quadrats (each quadrat of 0.25 m^2) and oven dried at 60 °C to constant weight. The shoot biomass of maize silage was sampled in late September before harvest. The plant samples were taken from 1 m^2. Plants were cut and oven dried at 60 °C to constant weight. To estimate root biomass, soil cores (15 cm depth and 5 cm in diameter) were taken. Roots were extracted from the soil samples by washing in nylon bags and dried at 60 °C. The content of plant N (PN), P (PP), and K (PK) in shoot biomass was determined as follows: the PN content was analyzed by the Kjeldahl method [AOAC], the content of PP was determined colorimetrically by a continuous flow SNA, and the PK content was analyzed by flame photometry. The amount of carbon in shoot and root biomass was calculated by multiplying the biomass by the conversion factor 0.475 [25].

2.3. Soil Loss Rate Calculation

To assess the annual soil loss rate, we used the Universal Soil Loss Equation (USLE) [26] as follows (Equation (3)):

$$A = R \times K \times LS \times C \times P \quad (3)$$

where A is the estimated average soil loss ($t.ha^{-1}.yr^{-1}$), R represents the rainfall erosivity factor ($MJ.mm^{-1}\ ha^{-1}.h^{-1}.yr^{-1}$), K represents the soil erodibility factor ($t.ha^{-1}$), SL represents the slope-steepness factor, C represents the crop/vegetation factor, and the support practice is expressed by P.

Rainfall erosivity factor (R) is determined (Equation (4)) as a sum of a storm kinetic energy event, EI30, multiplied by its maximum 30 min intensity, I30.

$$R = \frac{1}{N}\sum_{1}^{N}(Etot\ I30) \quad (4)$$

where R [$MJ.ha^{-1}cm.h^{-1}$] is a rainfall erosivity factor averaged for a period of length, N. Onderka and Pecho [27] calculated the mean annual R-factor for 95 Slovak localities and determined the mean annual value of the R-factor for Slovakia as 711.3 $MJ.ha^{-1}mmhr^{-1}$.

Soil erodibility factor (K) reflects soil sensitivity to erosion, and it is a function of soil properties, particularly soil texture and soil organic carbon matter as well. Based on soil texture and SOM content, the values of the K factor, following McKague and Eng [22], ranged from 0.04 to 0.43 for average SOM content, from 0.05 to 0.41 for SOM content less than 2%, and from 0.01 to 0.37 for SOM content more than 2%. In our study, the GL soil has a clay–loamy texture and SOM ranged from 31.42 $g.kg^{-1}$ to 48.62 $g.kg^{-1}$. The AL soil is loamy, and SOM ranged from 14.04 $g.kg^{-1}$ to 23.27 $g.kg^{-1}$. Based on these data, we determined the K factor for GL as 0.28 and for AL as 0.41.

The slope length–gradient factor (LS) defines the influence of topography on soil erosion. It describes a ratio of loss of soil under concrete conditions to that at a site with the standard slope steepness of 9% and slope length of 22.13 m. The risk of erosion increases with the longer and steeper slope. In the Kečovo locality, the length and steepness of the GL and AL slopes were 150 m and 10%, respectively. To calculate the LS factor, we used the equation that follows (Equation (5)):

$$\mathrm{LS} = [0.065 + 0.0456\ (\mathrm{slope}) + 0.006541\ (\mathrm{slope})^2](\frac{slope\ lenght}{constant})^{\mathrm{NN}} \quad (5)$$

where LS is the slope length–gradient factor, slope is sloping steepness in %, slope length is in m, constant is 22.1, and NN values dependent on the slope are 0.2 for slopes steeper less than 1%, 0.3 for slope steepness that ranged from 1% to 2.9%, 0.4 for slope steepness that ranged from 3% to 4.9%, and 0.5 for slopes steeper than 5%.

The crop/vegetation factor (C) is a ratio of soil loss from land under a specific crop and management system. The C factor combines a specific crop type and tillage method. The values for crop types are as follows: grain maize—0.40, silage maize, beans—0.50, cereals (spring and winer)—0.35, seasonal horticultural crops—0.50, fruit tress—0.10, and hay and pasture—0.02. The following values were determined for various tillage methods: fall plot—1.00, spring plough—0.90, mulch tillage—0.60, ridge tillage—0.35, zone tillage—0.25, and no-till—0.25. In our study, for the C factor for GL, we multiplied the crop type 0.02 for hay and pasture and 0.25 for the no-till tillage method. For the C factor for AL, we used a 0.5 value for silage maize and 0.6 value for the mulch tillage system.

Support practice factor (P) describes the influence of agricultural practice that will reduce the rate of soil erosion. The P factor reflects the ratio of soil loss by a support agricultural method to that of straight farming down and up the slope. Following McKague and Eng, the support practice values have been established as up to down practice—1.00, cross slope—0.75, contour farming—0.50, strip cropping, cross slope—0.37 and strip cropping, contour—0.25. Because for AL, strip cropping and cross slope practices were applied, we used a P factor of 0.37. For GL, the P factor was 0.25, as strip cropping and contour practices were applied.

2.4. Estimation of Soil Ecosystem Services Using Models

Soil water retention hydrological ecosystem services (WRHES) capacity, soil carbon accumulation ecosystem services (SCAES), and soil nitrogen accumulation ecosystem services (SNAES) were estimated using different models. We used different calculation formulas introduced by Gupta and Larson [28] for temperate pedotransfer functions [29] as field water capacity (FWC), wilting point (WP), and available water capacity (AWC) to estimate WRHES.

Field water capacity (FWC) (Equation (6)):

$$FWC = 0.003075 \times Sa + 0.005886 \times Si + 0.008039 \times Cl + 0.002208 \times SOM - 0.01434 \times BD \quad (6)$$

Wilting point (WP) (Equation (7)):

$$WP = 0.000059 \times Sa + 0.001142 \times Si + 0.0005766 \times Cl + 0.002228 \times SOM + 0.002671 \times BD \quad (7)$$

where *Sa* is the percentages of sand, *Si* is the percentage of silt particles, *Cl* is the percentage of clay particles, *SOM* is the percentage of soil organic matter, and *BD* is bulk density.

Available water capacity (AWC) (Equation (8)):

$$AWC = FWC - WP\ (\%) \tag{8}$$

where *FWC* is field water capacity in % and *WP* is the wilting point in %.

For the calculation of the soil water deficit index (SWDI) characterizing a drought event, the following formula was used (Equation (9)):

$$SWDI = \left(\frac{SMa - FWC}{AWC}\right) \times 10 \tag{9}$$

where *SMa* is actual soil moisture in %, *FWC* is field water capacity in %, and *AWC* is available water capacity in %. The SWDI was proposed by Martinez-Fernandez et al. [30]. We applied the SWDI as a proxy indicator to see differences between land use types and hillslope positions. The input data do not meet standard requirements (e.g., SMa assessment for one day only).

Higher values of SMa, FWC, WP, and AWC indicate higher soil water retention hydrological ecosystem services (WRHES) capacity. Positive SWDI values indicate that soils have excess water; when it equals zero, soil is in the field capacity of the water content (that is, without water deficit). Negative values mean soil drought, and the water deficit is absolute (wilting point) when the SWDI reaches ≤ -10 [31]. The SWDI in the range 0–−2 indicates a mild, −2–−5 moderate, −5–−10 severe, and ≤ -10 extreme drought level.

To estimate SCAES and SNAES, SOC stock (SOCS) (t.ha^{-1}) and SN stock (SNS) (t.ha^{-1}) were calculated according to Tan [32] and Chen [33] (Equation (10)):

$$SOCS = \sum_{i=1}^{n} SOCi \times \mathrm{BDi} \times \mathrm{Hi} \times \left(1 - \frac{\mathrm{G}}{100}\right) \times 10^{-1} \tag{10}$$

where SOCS is the SOC stock pool (t.ha^{-1}) of the evaluated soil profile, SOCi is SOC content (g.kg^{-1}), BDi is bulk density (g.cm^{-3}), Hi is the soil thickness (m), and G is the volume percent of gravel and stones (particle size > 2 mm) in layer i, respectively.

A similar approach was used to calculate SNS (t.ha^{-1})] (Equation (11)):

$$SNS = \sum_{i=1}^{n} SNi \times \mathrm{BDi} \times \mathrm{Hi} \times \left(1 - \frac{\mathrm{G}}{100}\right) \times 10^{-1} \tag{11}$$

where SNS is the total soil nitrogen stock (t.ha^{-1}) of the evaluated soil profile, SNi is total soil nitrogen content (g.kg^{-1}), BDi is bulk density (g.cm^{-3}), Hi is the soil thickness (m), and G is the volume percent of gravel and stones (particle size > 2 mm) in layer i, respectively.

2.5. Data Analysis

The data were analyzed by using the Pearson correlation coefficients and analysis of variance (ANOVA). The statistical analyses were computed using SPSS Statistics 28.

3. Results

3.1. Basic Soil Physical and Chemical Properties under Different Land Use, Hillslope Position, and Soil Depth

Table 1 presents basic soil physical properties in GL and AL at different hillslope positions and soil depths. In GL, the BD ranged from 1.41 (in the 0–10 cm depth at the lower hillslope position) to 1.55 g.cm^{-3} (in the 35–45 cm depth at the upper hillslope position). In AL, the BD ranged from 1.35 (in the 0–10 cm depth at the lower hillslope position) to 1.59 g.cm^{-3} (in the 0–10 cm depth at the middle hillslope position). At all three hillslope positions in GL, the higher BD values were recorded in the 35–45 cm depth. In contrary, at the upper and middle hillslope positions in AL, the higher BD values were recorded in the 0–10 cm depth. The PD ranged from 2.63 to 2.71 g.cm^{-3}. Soil texture was variable among the hillslope positions, depths, and land uses. Soil has a clay to clay–loamy texture in GL and a clay, loamy, and silty loamy texture in AL. In GL, fine clay fractions increased downward along the slope, and vice versa the sand fractions. On the contrary, in AL, coarse sand fractions increased downward along the slope, just like gravel. The SM values were always higher in the first soil layer and in GL. In the 0–10 cm soil depth, the lowest SM values were recorded in the upper hillslope position under both land uses. The highest SM value (40.8%) was recorded at the middle hillslope position in GL and at the lower hillslope position in AL (19.8%). Lower ST values were recorded in AL compared to GL.

Table 1. Soil physical properties.

Land Use	Hillslope Position	Soil Depth (cm)	BD ($g.cm^{-3}$)	PD ($g.cm^{-3}$)	Clay (%)	Silt (%)	Sand (%)	Gravel (%)	SM (%)	ST (°C)
Grassland	Upper	0–10	1.45 ± 0.05	2.69 ± 0.02	31.43 ± 2.36	24.15 ± 1.42	44.42 ± 3.50	2.3 ± 0.4	33.8 ± 6.06	14.5 ± 0.44
		35–45	1.55 ± 0.04	2.70 ± 0.01	39.58 ± 2.55	33.12 ± 2.21	27.30 ± 2.64	5.8 ± 1.0	19.3 ± 3.70	9.4 ± 0.48
	Middle	0–10	1.42 ± 0.02	2.70 ± 0.01	35.01 ± 1.26	23.11 ± 1.12	41.88 ± 0.92	5.9 ± 0.7	40.8 ± 2.28	13.8 ± 0.51
		35–45	1.48 ± 0.04	2.71 ± 0.01	40.56 ± 1.55	35.12 ± 1.07	24.32 ± 2.34	10.3 ± 0.5	23.8 ± 1.48	8.8 ± 0.45
	Lower	0–10	1.41 ± 0.02	2.70 ± 0.01	38.25 ± 1.75	28.14 ± 1.23	33.61 ± 1.71	5.7 ± 1.1	38.0 ± 1.87	15.0 ± 0.77
		35–45	1.52 ± 0.03	2.71 ± 0.01	43.78 ± 3.82	39.98 ± 1.38	16.24 ± 3.25	10.2 ± 0.5	21.0 ± 3.16	9.7 ± 0.13
Arable land	Upper	0–10	1.55 ± 0.11	2.64 ± 0.01	21.69 ± 3.88	54.98 ± 3.06	23.38 ± 4.69	5.1 ± 0.2	15.8 ± 1.15	10.9 ± 0.45
		35–45	1.51 ± 0.04	2.66 ± 0.01	46.23 ± 2.42	38.88 ± 1.73	14.89 ± 1.47	10.4 ± 0.5	9.0 ± 0.90	6.0 ± 0.25
	Middle	0–10	1.59 ± 0.05	2.65 ± 0.01	25.89 ± 2.43	47.95 ± 1.94	26.21 ± 2.67	20.3 ± 2.5	19.3 ± 1.17	10.4 ± 0.80
		35–45	1.58 ± 0.03	2.70 ± 0.03	48.97 ± 4.23	39.01 ± 3.61	12.02 ± 4.38	30.4 ± 3.1	11.8 ± 2.15	8.5 ± 0.55
	Lower	0–10	1.35 ± 0.06	2.63 ± 0.02	17.71 ± 2.78	53.15 ± 2.62	29.14 ± 3.44	30.5 ± 3.1	19.8 ± 1.78	9.9 ± 0.70
		35–45	1.53 ± 0.10	2.70 ± 0.09	16.78 ± 1.04	46.99 ± 1.01	36.23 ± 1.53	50.7 ± 3.9	14.8 ± 1.25	8.6 ± 0.40

BD—bulk density; PD—particle density; SM—soil moisture; ST—soil temperature.

Figure 2 presents penetration resistance (PR) with related penetration depths (D) in GL and AL at different hillslope positions. Significant differences were found among the hillslope positions and land uses with regard to PR and D.

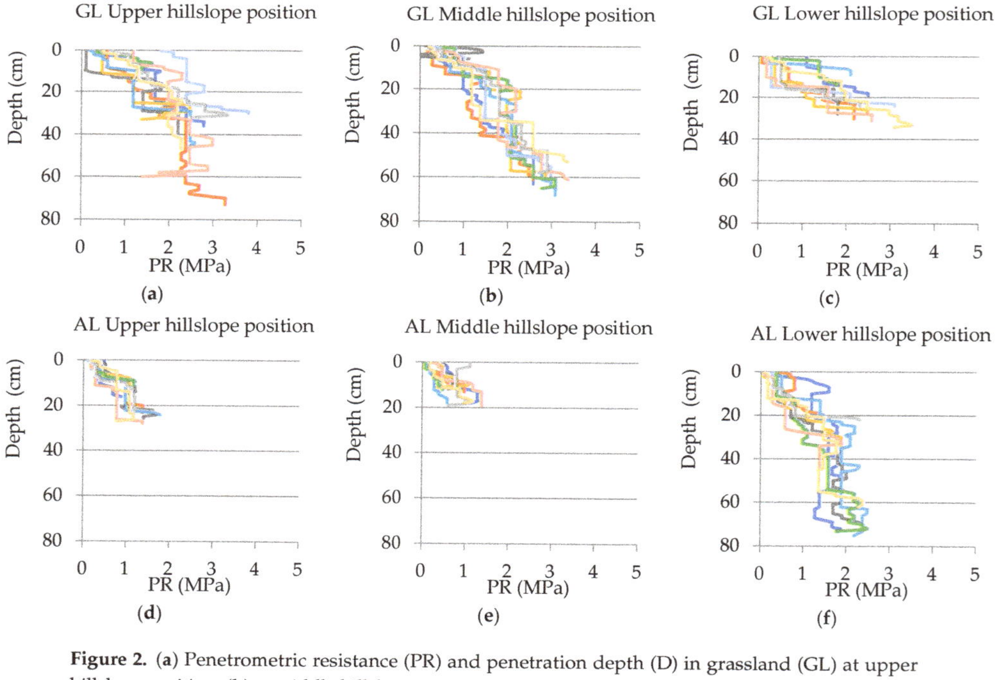

Figure 2. (**a**) Penetrometric resistance (PR) and penetration depth (D) in grassland (GL) at upper hillslope position; (**b**) at middle hillslope position; and (**c**) at lower hillslope position; (**d**) penetrometric resistance (PR) and penetration depth (D) in arable land (AL) at upper hillslope position; (**e**) at middle hillslope position; and (**f**) at lower hillslope position.

The mean PR for 20 cm depth ranged from 1.23 MPa at the upper hillslope position in GL to 0.67 MPa at the middle hillslope position in AL. The PR was significantly lower in AL compared to GL. In GL, the shallowest mean D (25 cm) was found at the lower hillslope position, and a much deeper mean D (42 cm) was measured at the upper and the deepest (60 cm) at the middle hillslope position, respectively. On contrary, in AL, the deepest mean D (52 cm) was found at the lower hillslope position as a consequence of accumulation processes which were reflected in the shallow mean D at the upper (24 cm) and middle (17 cm) hillslope positions.

Table 2 presents basic soil chemical properties in GL and AL at different hillslope positions and soil depths. In general, the soil reaction values ranged from weakly acidic, through acidic, to strongly acidic, from 4.78 to 5.93 in GL and from 4.85 to 5.54 in AL. The SOC content at the 0–10 cm depth ranged from 18.23 to 28.19 g.kg^{-1} in GL and from 8.15 to 13.50 g.kg^{-1} in AL. The SOC and SOM content was higher in the depth 0–10 cm compared to the 35–45 cm depth. However, the organic matter quality expressed by the SOC/SN ratio showed better quality in the 35–45 cm depth. The available SP content ranged from 0.15 to 22.83 mg.kg^{-1} in GL and from 4.33 to 24.91 in AL. The available SK content ranged from 84.40 to 349.3 mg.kg^{-1} in GL and from 130.75 to 221.93 in AL.

Table 2. Soil chemical properties.

Land Use	Hillslope Position	Soil Depth (cm)	pH	SOC (g.kg^{-1})	SOM (g.kg^{-1})	SN (g.kg^{-1})	SOC/SN	SP (mg.kg^{-1})	SK (mg.kg^{-1})
Grassland	Upper	0–10	5.04 ± 0.10	18.23 ± 2.81	31.42 ± 4.84	2.22 ± 0.29	8.21	0.15 ± 0.05	116.20 ± 5.31
		35–45	5.16 ± 0.07	6.98 ± 2.08	12.02 ± 3.59	1.19 ± 0.10	5.89	0.23 ± 0.08	84.40 ± 0.02
	Middle	0–10	5.14 ± 0.13	21.52 ± 2.98	37.09 ± 5.13	2.02 ± 0.20	10.66	1.28 ± 0.76	198.73 ± 47.02
		35–45	4.78 ± 0.28	7.25 ± 0.65	12.50 ± 1.12	1.05 ± 0.19	6.89	0.16 ± 0.03	99.84 ± 2.78
	Lower	0–10	5.89 ± 0.10	28.19 ± 2.10	48.62 ± 3.61	3.13 ± 0.29	9.01	22.83 ± 2.67	349.30 ± 3.50
		35–45	5.93 ± 0.15	24.15 ± 6.71	41.64 ± 11.58	2.75 ± 0.50	8.78	18.07 ± 8.50	321.33 ± 50.55
Arable land	Upper	0–10	5.22 ± 0.03	13.50 ± 3.31	23.27 ± 5.71	1.62 ± 0.09	8.33	24.91 ± 3.54	145.71 ± 20.67
		35–45	5.23 ± 0.07	13.05 ± 1.70	22.50 ± 2.93	1.58 ± 0.13	8.27	22.71 ± 2.50	130.75 ± 16.09
	Middle	0–10	4.85 ± 0.15	8.15 ± 0.77	14.04 ± 1.32	1.09 ± 0.10	7.47	5.12 ± 2.10	168.75 ± 29.79
		35–45	4.89 ± 0.12	7.73 ± 1.89	13.32 ± 3.27	1.04 ± 0.22	7.41	4.59 ± 3.41	179.65 ± 50.13
	Lower	0–10	5.46 ± 0.14	13.09 ± 3.00	22.56 ± 5.18	1.58 ± 0.09	8.31	4.84 ± 6.47	221.93 ± 86.20
		35–45	5.54 ± 0.08	8.63 ± 2.26	14.87 ± 3.90	1.42 ± 0.23	6.10	4.33 ± 6.70	182.87 ± 63.75

pH—soil reaction; SOC—soil organic carbon; SOM—soil organic matter; SN—soil total nitrogen; SP—soil plant-available phosphorus; SK—soil plant available potassium.

The values differed between hillslope positions and soil depths due to erosion–accumulation processes. The most significant accumulation processes at the lower hillslope position were manifested in the case of SK in GL as well as in AL. In GL, significant accumulation at the lower hillslope position was observed also in the case of SOC, SOM, SN, and SP. Overall, in PG, most chemical parameters (SK, SP, SOC, SOM, and pH) increased downward along the slope. In AL, most chemical parameters (pH, SOC, SOM, and SN) reached the lowest values at the middle hillslope position.

3.2. Grassland and Silage Maize Properties and Content of Nutrients with Different Hillslope Positions

Shoot biomass in GL ranged significantly among different hillslope positions, with the highest values in the lower part (223 g.m^{-2}) and the lowest in the middle parts of the hillslope positions (95 g.m^{-2}). The highest value of root biomass was recorded in the upper hillslope position (837 g.m^{-2}). Contrary to GL, the highest shoot biomass of maize (1445 g.m^{-2}) was obtained in the middle part of the hillslope, followed by the upper part and lower part (Table 3). On average, the root biomass in AL was 1.3 times lower than in GL, with the highest values in the upper part. R/S ratios were not related to hillslope position. In GL, the R/S ratio was largest in the middle hillslope part. Compared to GL, the largest R/S ratio in AL was observed in the upper part.

In GL, the content of macronutrients differed between hillslope positions (Table 4). In the lower part, the PN and PP concentrations were significantly higher ($p < 0.01$) compared to the upper part (Table 3). Grassland shoot biomass and PC/PN and PC/PP ratios decreased significantly downward along the slope with an increasing content of SN and

SP in the soil. To the contrary, an opposite trend was observed in AL, with the highest content of PN and PP and the lowest shoot PC/PN, PC/PP, and PN/PP ratios in the upper part, the lowest shoot PC/PN, PC/PP, and PN/PP ratios in the upper part. However, the content of macronutrients and PC/PN, PC/PP, PN/PP, and PN/PK ratios in maize did not exhibit significant change with the hillslope position.

Table 3. Shoot biomass, root biomass, and root biomass to shoot biomass ratio in grassland and maize grown on arable land.

	Hillslope Position	Shoot Biomass ($g.m^{-2}$)	Root Biomass ($g.m^{-2}$)	R/S
Grasslands	Upper	188 ± 88^{b}	837 ± 47^{ns}	5.23 ± 1.40^{ns}
	Middle	95 ± 35^{c}	663 ± 121^{ns}	6.57 ± 1.57^{ns}
	Lower	223 ± 59^{a}	611 ± 52^{ns}	2.61 ± 0.65^{ns}
	p value	0.009	0.179	0.107
Arable land	Upper	1186 ± 495^{ns}	340 ± 79^{a}	0.32 ± 0.14^{ns}
	Middle	1445 ± 448^{ns}	307 ± 24^{ab}	0.23 ± 0.08^{ns}
	Lower	1104 ± 593^{ns}	184 ± 45^{b}	0.21 ± 0.11^{ns}
	p value	0.826	0.003	0.171

Contrasting letters denote significant differences; ns—not significant, *p*—probability value.

Table 4. Content of macronutrients and PC/PN, PC/PP, PN/PP, and PN/PK ratios in grassland and maize grown on arable land.

	Hillslope Position	PN ($g.kg^{-1}$)	PP ($g.kg^{-1}$)	PK ($g.kg^{-1}$)	PC/PN	PC/PP	PN/PP	PN/PK
Grasslands	Upper	20.77 ± 1.16^{b}	2.97 ± 0.18^{b}	27.93 ± 3.10^{ns}	22.93 ± 1.29^{a}	160.41 ± 10.57^{a}	7.00 ± 0.40^{ns}	0.75 ± 0.11^{ns}
	Middle	21.56 ± 1.96^{b}	3.23 ± 0.21^{a}	22.43 ± 3.16^{ns}	22.20 ± 2.20^{a}	147.79 ± 10.92^{a}	6.68 ± 0.36^{ns}	0.97 ± 0.10^{ns}
	Lower	26.81 ± 2.60^{a}	4.02 ± 0.28^{a}	31.19 ± 3.28^{ns}	17.86 ± 1.65^{b}	118.56 ± 8.32^{b}	6.99 ± 0.36^{ns}	0.86 ± 0.06^{ns}
	p value	0.009	0.000	0.836	0.010	0.000	0.118	0.120
Arable land	Upper	14.66 ± 2.96^{ns}	4.38 ± 0.52^{ns}	7.64 ± 1.64^{ns}	33.68 ± 8.48^{ns}	109.61 ± 13.03^{b}	3.34 ± 0.57^{ns}	1.98 ± 0.62^{ns}
	Middle	13.66 ± 2.55^{ns}	3.44 ± 0.38^{ns}	9.83 ± 2.54^{ns}	35.74 ± 6.81^{ns}	139.07 ± 14.40^{a}	4.00 ± 0.97^{ns}	1.45 ± 0.34^{ns}
	Lower	11.81 ± 1.49^{ns}	3.45 ± 0.61^{ns}	7.71 ± 0.91^{ns}	40.71 ± 5.28^{ns}	141.35 ± 27.79^{a}	3.48 ± 0.51^{ns}	1.56 ± 0.32^{ns}
	p value	0.118	0.167	0.966	0.167	0.0472	0.804	0.213

PN—plant nitrogen; PP—plant phosphorous; PK—plant potassium; PC—plant carbon. Contrasting letters denote significant differences; ns—not significant, *p*—probability value.

3.3. *Relationships between Soil and Plant Properties*

The correlation coefficient matrix between soil and plant properties is presented in Table 5. The soil pH had a statistically significant positive relationship with other soil chemical properties, including plant nitrogen. The SOC had a significant positive correlation with all soil chemical properties, including two plant nutrients (PN and PK), and root biomass, while it had a significant negative correlation with shoot biomass, bulk density, and content of silt and gravel. For plant nutrients, the least numerous significant correlations with soil chemical and physical properties occurred in the case of PP. Soil physical properties affected mainly PK, ShB, and RoB.

3.4. *Annual Soil Loss by Water Erosion*

According to this study, the annual soil loss due to water erosion was strongly affected by land use. A negligible annual average soil loss was recorded for GL with permanent vegetation cover (0.76 $t.ha^{-1}.yr^{-1}$). To contrary, a very high rate of soil erosion was found for AL with maize silage (99.05 $t.ha^{-1}.yr^{-1}$). In the Slovak legislation [34], four soil severity classes are defined: low erosion severity (<4 $t.ha^{-1}.yr^{-1}$), moderate erosion (4–10 $t.ha^{-1}.yr^{-1}$), high erosion (10–30 $t.ha^{-1}.yr^{-1}$), and severe (>30 $t.ha^{-1}.yr^{-1}$). Based on the soil erosion severity category, the study site could be classified in the severe class.

Table 5. Correlation coefficients among soil and plant properties.

	pH	SOC	SN	SP	SK	PN	PP	PK	ShB	RoB	BD	PD	Clay	Silt	Sand	Gravel
pH	1															
SOC	0.587 **	1														
SN	0.682 **	0.950 **	1													
SP	0.477 **	0.490 **	0.490 **	1												
SK	0.700 **	0.714 **	0.720 **	0.528 **	1											
PN	0.457 *	0.780 **	0.797 **	N.C.	N.C.	1										
PP	N.C.	N.C.	N.C.	0.559 **	N.C.	N.C.	1									
PK	N.C.	0.730 **	0.769 **	N.C.	N.C.	0.869 **	N.C.	1								
ShB	N.C.	−0.689 **	−0.646 **	N.C.	N.C.	−0.653 **	N.C.	−0.788 **	1							
RoB	N.C.	0.521 **	0.466 *	N.C.	N.C.	0.726 **	N.C.	0.804 **	−0.680 **	1						
BD	N.C.	−0.374 **	−0.376 **	N.C.	N.C.	N.C.	N.C.	N.C.	N.C.	N.C.	1					
PD	N.C.	N.C.	N.C.	N.C.	N.C.	N.C.	N.C.	0.442 *	−0.414 *	N.C.	N.C.	1				
Clay	N.C.	N.C.	N.C.	N.C.	N.C.	0.808 **	N.C.	0.828 **	−0.677 **	0.637 **	N.C.	N.C.	1			
Silt	N.C.	−0.397 **	−0.386 **	N.C.	N.C.	−0.765 **	0.411*	−0.874 **	0.789 **	−0.775 **	N.C.	−0.338 *	−0.535 **	1		
Sand	N.C.	N.C.	N.C.	−0.361*	N.C.	0.523 **	−0.530 **	0.684 **	−0.683 **	0.696 **	−0.435 **	N.C.	−0.524 **	−0.439 **	1	
Gravel	N.C.	−0.410 **	−0.378 **	N.C.	N.C.	−0.658 **	N.C.	−0.637 **	0.568 **	−0.656 **	N.C.	N.C.	−0.406 **	0.509 **	N.C.	1

pH—soil reaction; SOC—soil organic carbon; SN—soil total nitrogen; SP—soil plant-available phosphorus; SK—soil plant available potassium; PN—plant nitrogen; PP—plant phosphorous; PK—plant potassium;, PC—plant carbon; ShB—shoot biomass; RoB—root biomass; BD—bulk density; PD—particle density; * r-values shown in bold are significant at $p < 0.05$; ** r-values shown in bold are significant at $p < 0.01$; N.C.—no correlation.

3.5. Potential of Soil Ecosystem Services Affected by Water Erosion

The impact of erosion–accumulation processes on soil properties was reflected in ecosystem services. Carbon and nitrogen stocks were influenced by hillslope positions, land use, and soil layer (Figure 3). Currently, the upper layer of soil is the most active in the ongoing carbon storage processes. However, deeper soil horizons also have a significant role in total carbon storage, which is obviously highlighted in the lower hillslope position in GL (SOCS 33.0 t.ha^{-1}). In GL, the SOCS for the top 10 cm layer ranged from 37.7 t.ha^{-1} in the lower hillslope position to 25.9 t.ha^{-1} in the upper hillslope position. The SOCS was significantly lower in AL, ranging from 19.9 t.ha^{-1} in the upper hillslope position to 10.4 t.ha^{-1} in the middle hillslope position. Similar trends were observed in the case of nitrogen stocks.

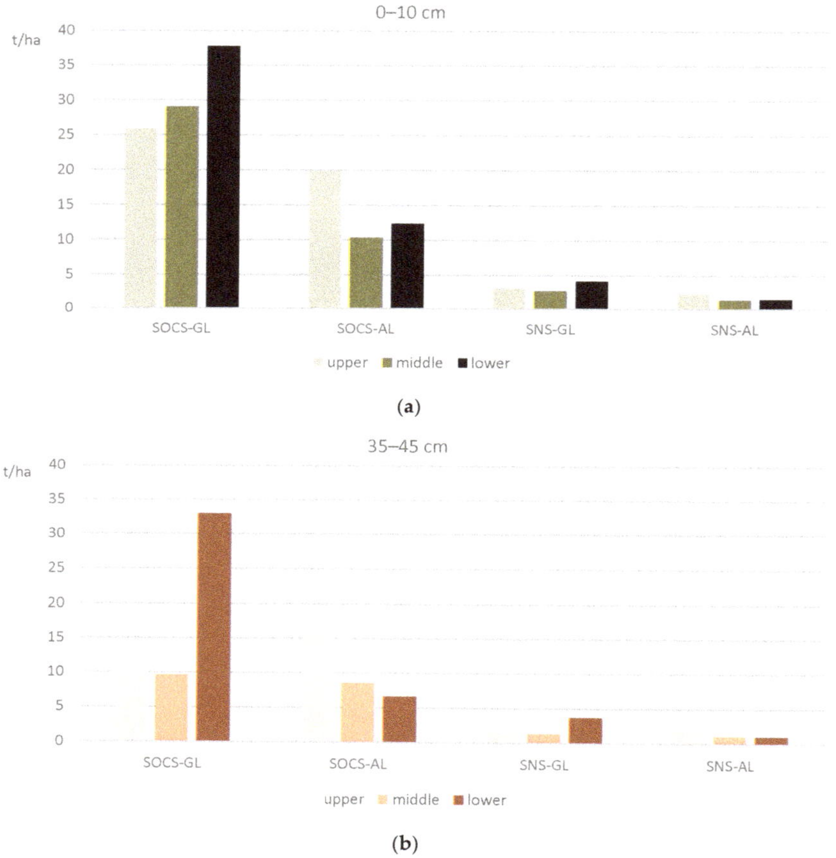

Figure 3. Soil organic carbon stock (SOCS) and soil total nitrogen stock (SNS) for 10 cm layers; (**a**) 0–10 cm depth; (**b**) 35–45 cm depth.

The effect of hillslope position, land use, and soil depth in the case of hydrological ecosystem services are obvious but not as pronounced as in the case of SOCS and SNS (Figure 4). The soil hydrological properties assessed by models using pedotransfer functions showed a much higher soil water retention potential in the case of WP in GL, and in the case of AWC in AL. However, the directly measured actual SM values were significantly higher in GL (37.5% as a mean value in the 0–10 cm depth) than in AL (18.3% as a mean value in the 0–10 cm depth), which was also reflected in the SWDI being more favorable for GL. The reason may be that the model does not consider the type of vegetation cover, which can contribute to increased water retention ecosystem services.

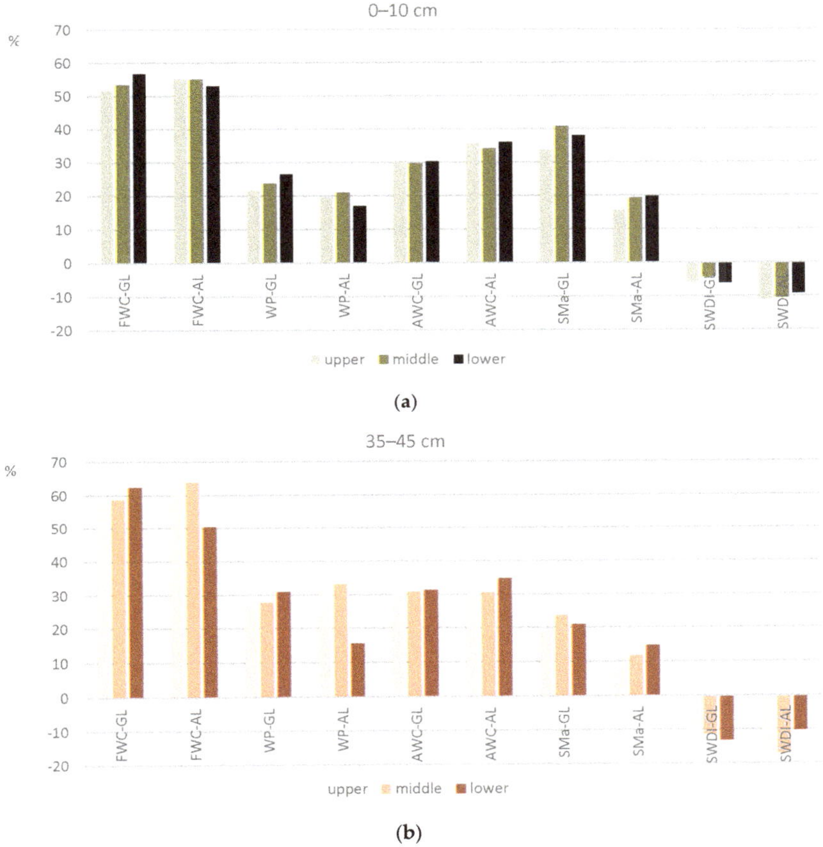

Figure 4. Field water capacity (FWC), wilting point (WP), available water capacity (AWC), actual soil moisture (SMa), soil water deficit index (SWDI); (**a**) 0–10 cm depth; (**b**) 35–45 cm depth.

4. Discussion

4.1. Soil and Plant Properties Affected by Water Erosion

We recorded a significant change in soil and plant properties in different hillslope positions, with an impact on the functioning of ecosystem services.

For the soil physical properties, significant changes along the slope were recorded in the values of penetrometric resistance as well as the depth of its measurement. Slope, pasturing, and type of vegetation strongly influence soil physical properties, including pentrometric resistance [35]. Soil resistance is one of the most dynamic properties of soil and it is important for plant growth and soil biological activities [36]. Lower PR at 20 cm was recorded in AL compared to GL as a consequence of ploughing. Better PR values in AL were in contrast to penetrometric depth, which was the shallowest (17 cm) at the middle hillslope position, where in GL, on the contrary, the greatest depth of penetrometric measurements was recorded (60 cm). The deepest penetrometric depth in AL was recorded at the lower hillslope position (52 cm) as a consequence of accumulation processes. This confirms how much impact erosion has on soil depth, which is largely corrected by vegetation and management. Erosion can reduce soil thickness and transport soil particles, and thus impact other soil properties, including water availability and thus soil hydrological ecosystem services [37].

Soil particles and their size and density are the main drivers of the transport and sedimentation process [38–40]. Previous studies showed that, as a rainy season lengthens, the eroded sediment particles become coarser and more stable [41]. Their distribution due to erosion along the slope was different between GL and AL. In AL, coarse sand fractions increased upward along the slope, just like gravel. On the contrary, in GL, fine clay fractions increased downward along the slope. Similarly, but in cropland, Liu et al. [42] recorded an elevated percentage of clay and silt content at the bottom slope position after soil redistribution was induced by erosion. Moreover, they also found mutual interactions between size fractions and SOC content. The high quantity of fine particles at the bottom-slope position was accompanied by an increase in SOC content, just like we noticed in GL.

The transport and loss of soil particles, together with weathering, have an impact on soil fertility [43] and nutrient (C, N, P, and K) supplies [44]. On the other hand, erosion can influence nutrient losses [45,46], and their spatial distribution. In GL, nutrients, such as SK and SP, increased downward along the slope, with the highest concentration at the lower hillslope position, together with SN. An increase of SN along a decreasing linear transect was shown in studies on eroded natural and artificial grasslands in China [47,48]. In AL, the content of SN reached the lowest values at the middle hillslope position with the shallowest depth. The content of SP and SK showed different variation. While the concentration of SK increased downward along the slope, the content of SP was 20% lower at the lower hillslope position compared to the upper position. This very low SP content could be caused by the very low downwards mobility of phosphorus [49]. The same tendency of a higher soil nutrient content in the lower part of the slope was also manifested in the highest content of PN, PP, and PK in shoot biomass for GL. Similarly, in AL, the highest content of soil SN and SP was reflected in the highest content in maize shoot biomass in the upper part of the slope. The soil nutrient redistribution with the water erosion also induced changes in plant nutrient dynamics and stoichiometry [50]. In our study, higher availability of SN was positively correlated with PN ($r = 0.797^{**}$, $p < 0.01$) and PK ($r = 0.769^{**}$, $p < 0.01$). Previous studies showed that a higher content of soil nutrients, particularly in grasslands [51], is associated with a higher content of PN, PP, and PK and lower shoot PC/PN and PC/PP ratios [52]. A higher availability of SN and a low PC/PN ratio results in decreased nitrogen use efficiency in terrestrial ecosystems [53]. The PN/PP and PN/PK ratios show how the ecosystem is limited by nitrogen, potassium, and phosphorus. As the PN/PP ratio was lower than 14 [54], and PN/PK was lower than 2.1 [55], the shoot biomass of grassland and silage maize were nitrogen limited in upper, middle, and lower hill positions as well.

4.2. Ecosystem Services Affected by Water Erosion

Soil properties condition the potential of ecosystems to provide services [56]. Therefore, to sustain the capacity to supply ecosystem services, soil protection is crucial [57]. The evaluated properties affect the potential of supporting, provisioning, and regulating ecosystem services.

Soil loss caused by erosion significantly disrupts soil formation processes as an important supporting service. In general, grasslands are more effective at controlling soil erosion than other vegetation covers. Because of a lower susceptibility to soil erosion [58], the maintenance of permanent grasslands and introducing temporary leys as green cover have been some of the priorities of Common Agricultural Policy since 2014 [59]. At our GL study site, annual average soil loss reached 0.76 t/ha/yr, which is a relatively negligible soil erosion rate without a major impact on soil ecosystem services [60]. To the contrary, AL with a silage maize cropping system showed extreme severe soil loss. Severe erosion on arable land in East Slovakia was shown by Koco et al. [61], who estimated soil loss of 18.1 t/ha for spring barley over a period of four months from March 2021 to July 2021 using remote sensing techniques.

Soil erosion is reported to have a significant effect on crop yield and thus on provisioning ecosystem services. In GL, a significantly lower shoot biomass was recorded in the middle hillslope position and in the shallowest soil depth compared to the lower position. Similarly, a study on the impact of temporary and permanent grasslands on slope stability in Lithuania [62] showed that grassland production in the southern exposition was the highest at the foot slope because of its more favorable soil moisture content. In AL, silage maize did not show a reduction rate of shoot biomass yield. These results contrast with a study from China where the mean maize seed yields decreased by 35% for severe erosion sites [63]. However, a European study on crop production showed that productivity in Northern Europe with current land use and soil loss is not significantly lowered due to water soil erosion. Nevertheless, in the Mediterranean region, yield reduction induced by soil erosion is stronger than in Northern Europe [64].

Soil erosion and spatial pattern conversion of land use results into the changes of regulating ecosystem services, such as carbon and nitrogen stocks or water-related ecosystem services [65], which we observed at our study sites. The soil hydrological properties displayed a much greater soil water retention potential in the case of WP in GL, and in the case of AWC in AL. However, the directly field-measured actual SM values were significantly higher in GL compared to AL. Grasslands are considered important in providing water retention as an ecosystem service [66]. However, GL degradation usually has an effect on the deterioration of soil hydraulic properties and, thus, soil water retention [67].

Globally, soils are considered as the largest terrestrial carbon reservoir, with a stock of approximately 2135 Gt SOC [68]. GL soils store about 10% of the global SOCS, which is nearly 50% more than in forest soils [69]. GL SOCSs are usually higher compared to AL SOCSs [70], as was confirmed by the results in our study, because of more plant-derived C inputs, mainly coming from root biomass and permanent cover [71]. In addition, the SOCS in AL is still being lost under current agricultural management [72], which is accelerated under erosion responsible for the degradation of physical and chemical soil properties. Our results showed the impact of erosion on C runoff, with subsequent redistribution of SOCSs within the soil profile under various land use. The land use and the intensity of management showed a difference in the SOC stocks as well as in their distribution at the different hillslope positions at the study sites. In GL, significantly higher SOC stocks were recorded compared to AL. However, in GL, we recorded the highest stocks and thus the accumulation of SOC at the lower hillslope position, in both monitored depths, with a greater emphasis in the second depth of 35–45 cm. In AL, we recorded the highest SOCS in the upper part of the slope, at a depth of 0–10 cm, and the lowest in the middle part of the slope. These variations in SOC reserves are a reflection of several linked biogeochemical processes taking place together with erosion processes and simultaneously influencing each other. In this way, the supply and distribution of nutrients in different parts of the slope are influenced, thus generating the formation of biomass as a prerequisite for SOC supplies. Carbon transfer as a result of erosion processes can control the global carbon cycle, with consequences on the ecosystem services and the evolution of the climate. Erosion induces a transfer of organic carbon from vegetation to river and sediments, and thereby acting to remove atmospheric CO_2 jointly with silicate weathering [7]. In our study, SOC was not transferred to a river, but was dominantly accumulated at the lower hillslope position. This presupposes the formation of colluvial soils. Soil type is a phenomenon as important as land use or other factors, having an impact on carbon stocks because of differences in geological material and the degree of its breakdown. Globally, large carbon stocks are in Cambisols, due to their vast area they cover [73]. Cambisols are relatively young soils with incipient subsoil formation connected with Fe oxides and clay mineral formation pedogenic processes. They predominantly accumulate organic matter in a top soil horizon. SOC content and then SOC stocks are further affected by other soil properties, as shown by correlation analyses.

Soil water erosion is a natural phenomenon and its negative impact on soil ecosystem services can be significantly reduced by sustainable land use and soil conservation management. The use of cover crops in areas with high soil erosion potential combined with limited soil disturbance technologies can partially or completely reduce the impact of climate change on soil losses [74]. Agri-environmental measures incorporated into the EU's Common Agricultural Policy (CAP) are considered to be the most efficient tool to stop soil degradation by erosion in the EU. Generally, more attention should be given to grasslands in erosion-regulating practices and ecosystem services. Much ecosystem services research is currently focused on forests and wetlands [75] or arable lands.

5. Conclusions

This study confirmed the effect of soil erosion on soil physical, chemical, and plant properties and related ecosystem services under different land uses.

The transfer of soil material from the upper and middle part of the slope and its accumulation in the lower part was evident and confirmed at AL by soil depth measurement, and sand particles increased downward along the slope. In contrast to AL, at GL, the shallowest profile was found at the lower hillslope position, and finest clay fractions increased downward along the slope. Overall, in PG, most soil chemical parameters (SK, SP, SOC, SOM, and pH) increased downward along the slope, where plant biomass and some plant nutrient concentrations (PN and PP) were also significantly higher compared to the upper part. However, in contrast, root biomass displayed the highest amount in the upper hillslope position. The content of macronutrients in maize did not exhibit significant changes with the hillslope position, probably due to fertilizer application. Correlation analysis showed the relations between soil and plant properties. A significant positive correlation was found between SOC and all soil chemical properties, including two plant nutrients (PN and PK), and root biomass. For plant nutrients, the least numerous significant correlations with soil chemical and physical properties occurred in the case of PP. Soil physical properties affected mainly PK, ShB, and RoB.

Soil and plant properties were reflected in ecosystem services. The study displayed that the annual soil loss by water erosion was strongly affected by land use. A relatively negligible annual average soil loss was recorded for GL. To the contrary, a very high rate of soil erosion was found for AL with maize silage. The upper layer of soil was the most active in the ongoing carbon storage processes. However, deeper soil horizons have also a significant role in total carbon storage, which was highlighted in the lower hillslope position in GL. The SOCS was significantly lower in AL compared to GL. Similar trends were observed in the case of nitrogen stocks. The differences in hydrological ecosystem services were not so pronounced between different land uses. The estimated soil hydrological properties showed a much higher soil water retention potential in the case of WP in GL and in the case of AWC in AL.

The results showed how serious changes are taking place in the soils affected by erosion, with consequences on plants and ecosystem services. Hilly and mountainous areas require even more careful and complex agricultural management than lowland areas, due to soil erosion problems. The conservation cropping systems for hills and mountains should be based on favoring permanent crops such as grasslands, the expansion of rotation, the use of cover crops, and the reduction of tillage. These interventions will have a significant effect on the agricultural ecosystems of Slovakia and of many European environments.

Author Contributions: Conceptualization, R.K. and M.K.; methodology, M.K.; formal analysis, R.K., M.K., Ľ.J., J.Č. and Z.D.; investigation, R.K. and M.K.; data curation, R.K.; writing—original draft preparation, R.K. and M.K.; supervision, R.K.; project administration, M.K.; funding acquisition, M.K. All authors have read and agreed to the published version of the manuscript.

Funding: This research was funded by Ministry of Agriculture and Rural Development of the Slovak Republic, grant number 720/2023/MPRVSR–930.

Institutional Review Board Statement: Not applicable.

Informed Consent Statement: Not applicable.

Data Availability Statement: Data are contained within the article or are available from the corresponding author on request.

Conflicts of Interest: The authors declare no conflicts of interest.

References

1. Borrelli, P.; Robinson, D.A.; Fleischer., L.R.; Lugato, E.; Ballabio, C.; Alewell, C.; Meusburger, K.; Modugno, S.; Schütt, B.; Ferro, V.; et al. An assessment of the global impact of 21st century land use change on soil erosion. *Nat. Commun.* **2017**, *8*, 2013. [CrossRef] [PubMed]
2. Tuo, D.; Xu, M.; Gao, L.; Zhang, S.; Liu, S. Changed surface roughness by wind erosion accelerates water erosion. *J. Soils Sediments* **2016**, *16*, 105–114. [CrossRef]
3. Swezey, S.C.; Fitzwater, B.A.; Whittecar, G.R.; Mahan, A.A.; Garrity, C.P.; González, W.B.A.; Dobbs, K.M. The Carolina Sandhills: Quaternary eolian sand sheets and dunes along the updip margin of the Atlantic Coastal Plain province, southeastern United States. *Quat. Res.* **2016**, *86*, 271–286. [CrossRef]
4. Nearing, M.A.; Xie, Y.; Liu, B.; Yu, Y. Natural and anthropogenic rates of soil erosion. *Int. Soil Water Conserv. Res.* **2017**, *5*, 77–84. [CrossRef]
5. Keesstra, S.D.; Bouma, J.; Wallinga, P.T.; Smith, P.; Cerda, A.; Montanarella, L.; Quinton, J.N.; Pachepsky, Y.; van der Putten, W.H.; Bardgett, R.; et al. The significance of soils and soil science towards realization of the United Nations Sustainable Development Goals. *Soil* **2016**, *2*, 111–128. [CrossRef]
6. Guerra, A.J.T.; Fullen, M.A.; Jorge, M.d.C.O.; Bezerra, J.F.R.; Shokr, M.S. Slope processes, mass movement and soil erosion: A review. *Pedosphere* **2017**, *27*, 27–41. [CrossRef]
7. Hilton, R.G.; West, A.J. Mountains, erosion and the carbon cycle. *Nat. Rev. Earth Environ.* **2020**, *1*, 284–299. [CrossRef]
8. Chalise, D.; Kumar, L.; Kristiansen, P. Land degradation by soil erosion in Nepal: A Review. *Soil. Syst.* **2019**, *3*, 12. [CrossRef]
9. Ballabio, C.; Borrelli, P.; Spinoni, J.; Meusburger, K.; Michaelides, S.; Beguería, S.; Klik, A.; Petan, S.; Janeček, M.; Olsen, P.; et al. Mapping monthly rainfall erosivity in Europe. *Sci. Total Environ.* **2017**, *79*, 1298–1315. [CrossRef]
10. Poręba, G.J.; Śnieszko, Z.; Moska, P. Influence of pedon history and washing nature on luminescence dating of Holocene colluvium on the example of research on the Polish loess areas. *Quat. Int.* **2013**, *296*, 61–67. [CrossRef]
11. Porder, S.; Johnson, A.H.; Xing, H.X.; Brocard, G.; Goldsmith, S.; Pett-Ridge, J. Linking geomorphology, weathering and cation availability in the Luquillo Mountains of Puerto Rico. *Geoderma* **2015**, *249*, 100–110. [CrossRef]
12. Halecki, W.; Kruk, E.; Ryczek, M. Evaluation of water erosion at a mountain catchment in Poland using the G2 model. *Catena* **2018**, *164*, 116–124. [CrossRef]
13. Thaler, E.A.; Larsen, I.J.; Yu, Q. The extent of soil loss across the US Corn Belt. *Proc. Natl. Acad. Sci. USA* **2020**, *118*, e1922375118. [CrossRef] [PubMed]
14. Vilček, J.; Koco, Š. Integrated index of agricultural soil quality in Slovakia. *J. Maps* **2018**, *14*, 68–76. [CrossRef]
15. Kobza, J.; Barančíková, G.; Makovníková, J.; Pálka, B.; Styk, J.; Širáň, M. Current state and development of land degradation processes based on soil monitoring in Slovakia. *Agriculture* **2017**, *63*, 74–85. [CrossRef]
16. Gessert, A. Geomorphology of the Slovak Karst (Eastern Part). *J. Maps* **2016**, *12*, 285–288. [CrossRef]
17. Mello, J.; Elečko, M.; Pristaš, J.; Reichwalder, P.; Snopko, L.; Vass, D.; Steiner, A. *A Legend to the Geological Map of the Slovak Karst in 1:50,000*; Geological Institute of Dionýz Štúr: Bratislava, Slovakia, 1997; 255p. (In Slovak)
18. IUSS Working Group WRB. World Reference Base for Soil Resources. In *International Soil Classification System for Naming Soils and Creating Legends for Soil Maps*, 4th ed.; International Union of Soil Sciences (IUSS): Vienna, Austria, 2022; Available online: https://www.isric.org/sites/default/files/WRB_fourth_edition_2022-12-18.pdf (accessed on 20 June 2024).
19. Kobza, J.; Barančíková, G.; Dodok, R.; Hrivňáková, K.; Makovníková, J.; Styk, J.; Širáň, M. *Monitoring and Evaluation of Soil Properties in Slovakia and the Potential of Their Development*; Soil Science and Conservation Research Institute: Bratislava, Slovakia, 2013; 184p. (In Slovak)
20. Klika, J.; Novák, A.; Gregor, A. *Practical Lessons in Phytocenology, Ecology, Climatology and Soil Science*; NCSAV: Praha, Czech Republic, 1954. (In Czech)
21. Blake, G.R.; Hartage, K.H. Particle density. In *Methods of Soil Analysis, Part 1, Physical and Mineralogical Methods—Agronomy Monograph No.9*; American Society of Agronomy—Soil Science Society of America: Madison, WI, USA, 1986; pp. 377–382.
22. Nikitin, V.; Fishman, V. On the improvement of methods for determination of soil carbon. *Chem. Agric.* **1969**, *3*, 76–77.
23. AOAC. *Official Methods of Analysis*, 17th ed.; AOAC: Washington, DC, USA, 2000.
24. Mehlich, A. Mehlich 3 soil test extractant: A modification of Mehlich 2 extractant. *Commun. Soil Sci. Plant Anal.* **1984**, *15*, 1409–1416. [CrossRef]
25. Schlesinger, W.H. *Biochemistry: An Analysis of Global Change*; Academic Press: San Diego, CA, USA, 1991; 432p, ISBN 978-0-12-625157-9.
26. McKague, K.; Eng, P.; OMAFRA Factsheet. Universal Soil Loss Equation (USLE). Ontario's Ministry of Agriculture, Food and Agribusiness and Ministry of Rural Affairs 2023. Available online: https://files.ontario.ca/omafra-universal-soil-loss-equation-23-005-en-2023-03-02.pdf (accessed on 10 June 2024).

27. Onderka, M.; Pecho, J. Update of the erosive rain factor in Slovakia using data from the period 1961–2009. *Contrib. Geophys. Geod.* **2019**, *49*, 355–371. [CrossRef]
28. Gupta, S.C.; Larson, W.E. Estimating soil water retention characteristics from particle size distribution, organic matter percent, and bulk density. *Water Resour. Res.* **1979**, *15*, 1633–1635. [CrossRef]
29. Botula, Y.D.; Cornelis, W.M.; Baert, G.; Van Ranst, E. Evaluation of pedotransfer functions for predicting water retention of soils in Lower Congo (D.R.Congo). *Agric. Water Manag.* **2012**, *111*, 1–10. [CrossRef]
30. Martínez-Fernández, J.; González-Zamora, A.; Sánchez, N.; Gumuzzio, A. A soil water based index as a suitable agricultural drought indicator. *J. Hydrol.* **2015**, *522*, 265–273. [CrossRef]
31. Sevage, M.J.; Ritchie, J.T.; Bland, W.L.; Dugas, W.A. Lower limit of soil water availability. *Agron. J.* **1996**, *88*, 644–651. [CrossRef]
32. Tan, Q.; Wang, G.; Liu, X.; Hao, T.; Tan, W. Responses of soil organic carbon turnover to nitrogen deposition are associated with nitrogen input rates: Derived from soil ^{14}C evidences. *Environ. Pollut.* **2018**, *238*, 500–507. [CrossRef] [PubMed]
33. Chen, H.; Li, D.; Gurmesa, G.A.; Yu, G.; Li, L.; Zhang, W.; Fang, H.; Mo, J. Effects of nitrogen deposition on carbon cycle in terrestrial ecosystems of China: A meta-analysis. *Environ. Pollut.* **2015**, *206*, 352–360. [CrossRef]
34. Decree of the Ministry of Agriculture and Rural Development of the Slovak Republic No. 59 of 2013 on the Protection and Usage of Agricultural Land (In Slovak). Available online: https://www.slov-lex.sk/pravne-predpisy/SK/ZZ/2013/59/ (accessed on 15 June 2024).
35. Bayat, H.; Sheklabadi, M.; Moradhaseli, M.; Ebrahimi, E. Effects of slope aspect, grazing, and sampling position on the soil penetration resistance curve. *Geoderma* **2017**, *303*, 150–164. [CrossRef]
36. Tahmasbi, M.; Hemmat, A.; Vafaian, M.; Masaddeghi, M. Evaluation of Soil Compaction Strength (Pre-compaction Stress) Using Plate Sinkage and Uniaxial Confined Compression Tests. *JWSS-J. Water Soil Sci.* **2008**, *12*, 245–254. Available online: http://jstnar.iut.ac.ir/article-1-883-en.html (accessed on 15 June 2024).
37. Milodowski, D.T.; Mudd, S.M.; Mitchard, E.T.A. Erosion rates as a potential bottom-up control of forest structural characteristics in the Sierra Nevada Mountains. *Ecology* **2015**, *96*, 31–38. [CrossRef]
38. Asadi, H.; Ghadiri, H.; Rose, C.W.; Yu, B.; Hussein, J. An investigation of flow-driven soil erosion processes at low stream powers. *J. Hydrol.* **2007**, *342*, 134–142. [CrossRef]
39. Lima, J.L.M.P.; Dinis, P.A.; Souza, C.S.; Lima, M.I.P.; Cunha, P.P.; Azevedo, J.M.; Singh, V.P.; Abreu, J.M. Patterns of grain-size temporal variation of sediment transported by overland flow associated with moving storms: Interpreting soil flume experiments. *Nat. Hazards Earth Syst. Sci.* **2011**, *11*, 2605–2615. [CrossRef]
40. Zhang, P.; Yao, W.; Liu, G.; Xiao, P.; Sun, W. Experimental study of sediment transport processes and size selectivity of eroded sediment on steep Pisha sandstone slopes. *Geomorphology* **2020**, *363*, 107211. [CrossRef]
41. Grangeon, T.; Legout, C.; Esteves, M.; Gratiot, N.; Navratil, O. Variability of the particle size of suspended sediment during highly concentrated flood events in a small mountainous catchment. *J. Soils Sediments* **2012**, *2*, 1549–1558. [CrossRef]
42. Zhu, Y.; Wang, D.; Wang, X.; Li, W.; Shi, P. Aggregate-associated soil organic carbon dynamics as affected by erosion and deposition along contrasting hillslopes in the Chinese Corn Belt. *Catena* **2021**, *199*, 105106. [CrossRef]
43. Martínez-Mena, M.; Carrillo-L'épez, E.; Boix-Fayos, C.; Almagro, M.; García Franco, N.; Díaz-Pereira, E.; Montoya, I.; de Vente, J. Long-term effectiveness of sustainable land management practices to control runoff, soil erosion, and nutrient loss and the role of rainfall intensity in Mediterranean rainfed agroecosystems. *Catena* **2020**, *187*, 104352. [CrossRef]
44. Morford, S.L.; Houlton, B.Z.; Dahlgren, R.A. Geochemical and tectonic uplift controls on rock nitrogen inputs across terrestrial ecosystems. *Glob. Biogeochem. Cycles* **2016**, *30*, 333–349. [CrossRef]
45. Hilton, R.G.; Galy, A.; West, A.J.; Hovius, N.; Roberts, G.G. Geomorphic control on the δ15N of mountain forests. *Biogeosciences* **2013**, *10*, 1693–1705. [CrossRef]
46. Weintraub, S.R.; Taylor, P.G.; Porder, S.; Cleveland, C.C.; Asner, G.P.; Townsend, A.R. Topographic controls on soil nitrogen availability in a lowland tropical forest. *Ecology* **2015**, *96*, 1561–1574. [CrossRef]
47. Guoxiao, W.; Wang, Y. Using 137Cs to quantify the redistribution of soil organic carbon and total N affected by intensive soil erosion in the headwaters of the Yangtze River, China. *Appl. Radiat. Isot.* **2008**, *66*, 2007–2012. [CrossRef] [PubMed]
48. Li, Z.; Liu, C.; Dong, Y.; Chang, X.; Nie, X.; Liu, L.; Haibing, X.; Yinmei, L.; Zeng, G. Response of soil organic carbon and nitrogen stocks to soil erosion and land use types in the Loess hilly–gully region of China. *Soil Tillage Res.* **2017**, *166*, 1–9. [CrossRef]
49. Bertol, I.; Engel, F.L.; Mafra, A.L.; Bertol, O.J.; Ritter, S.R. Phosphorus, potassium and organic carbon concentrations in runoff water and sediments under different soil tillage systems during soybean growth. *Soil Tillage Res.* **2007**, *94*, 142–150. [CrossRef]
50. Holz, M.; Augustin, J. Erosion effects on soil carbon and nitrogen dynamics on cultivated slopes: A meta-analysis. *Geoderma* **2021**, *397*, 115045. [CrossRef]
51. Luo, W.; Li, M.H.; Sardans, J.; Lü, X.T.; Wang, C.; Peñuelas, J.; Wang, Z.; Han, X.G.; Jiang, Y. Carbon and nitrogen allocation shifts in plants and soils along aridity and fertility gradients in grasslands of China. *Ecol. Evol.* **2017**, *7*, 6927–6934. [CrossRef] [PubMed]
52. Yang, Y.; Luo, Y.; Lu, M.; Schädel, C.; Han, W. Terrestrial C:N stoichiometry in response to elevated CO_2 and N addition: A synthesis of two meta-analyses. *Plant Soil* **2011**, *343*, 393–400. [CrossRef]
53. Vitousek, P.M.; Porder, S.; Houlton, B.Z.; Chadwick, O.A. Terrestrial phosphorus limitation: Mechanisms, implications, and nitrogen-phosphorus interactions. *Ecol. Appl.* **2010**, *20*, 5–15. [CrossRef] [PubMed]
54. Gũsewell, S. N: P ratios in terrestrial plants: Variation and functional significance. *New Phytol.* **2004**, *164*, 243–266. [CrossRef]

55. OldeVentrik, H.; Wassen, M.J.; Verkroost, A.W.M.; De Ruiter, P.C. Species richness-productivity patterns differ between N-, P-, and K-limited wetlands. *Ecology* **2003**, *84*, 2191–2199. [CrossRef]
56. Dominati, E.; Mackay, A.; Green, S.; Patterson, M. A soil change-based methodology for the quantification and valuation of ecosystem services from agro-ecosystems: A case study of pastoral agriculture in New Zealand. *Ecol. Econ.* **2014**, *100*, 119–129. [CrossRef]
57. Steinhoff-Knopp, B.; Kuhn, T.K.; Burkhard, B. The impact of soil erosion on soil-related ecosystem services: Development and testing a scenario-based assessment approach. *Environ. Monit. Assess.* **2021**, *193*, 274. [CrossRef]
58. Milazzo, F.; Francksen, R.M.; Zavattaro, L.; Abdalla, M.; Hejduk, S.; Enri, S.R.; Pittarello, M.; Price, P.N.; Schils, R.L.; Smith, P.; et al. The role of grassland for erosion and flood mitigation in Europe: A meta-analysis. *Agric. Ecosyst. Environ.* **2023**, *348*, 108443. [CrossRef]
59. Panagos, P.; Ballabio, C.; Poesen, J.; Lugato, E.; Scarpa, S.; Montanarella, L.; Borrelli, P. A Soil Erosion Indicator for Supporting Agricultural, Environmental and Climate Policies in the European Union. *Remote Sens.* **2020**, *12*, 1365. [CrossRef]
60. Ruysschaert, G.J.; Poesen, G.; Verstraeten, G.; Govers, G. Soil loss due to harvesting of various crop types in contrasting agro-ecological environments. *Agric. Ecosyst. Environ.* **2007**, *120*, 153–165. [CrossRef]
61. Koco, Š.; Vilček, J.; Torma, S. Alternative Mapping Methods of Water Erosion. Agroporadenstvo 2023. Available online: https://www.agroporadenstvo.sk/nove-poznatky-poda?article=3047 (accessed on 15 July 2024).
62. Skuodienė, R.; Kinderienė, I.; Tomchuk, D.; Šlepetys, J. Root development of temporary and permanent grasslands and their anti-erosion significance on a hilly terrain. *Zemdirb.-Agric.* **2020**, *107*, 209–216. [CrossRef]
63. Duan, X.; Liu, B.; Gu, Z.; Rong, L.; Feng, D. Quantifying soil erosion effects on soil productivity in the dry-hot valley, southwestern China. *Environ. Earth Sci.* **2016**, *75*, 1164. [CrossRef]
64. Bakker, M.M.; Govers, G.; Jones, R.A.; Rounsevell, M.D. The effect of soil erosion on Europe's crop yields. *Ecosystems* **2007**, *10*, 1209–1219. [CrossRef]
65. Liang, J.; Li, S.; Li, X.; Liu, Q.; Meng, Q.; Lin, A.; Li, J. Trade-off analyses and optimization of water-related ecosystem services (WRESs) based on land use change in a typical agricultural watershed, southern China. *J. Clean. Prod.* **2021**, *279*, 123851. [CrossRef]
66. Trnka, M.; Balek, J.; Semenov, M.A.; Semerádová, D.; Bělínová, M.; Hlavinka, P.; Olesen, J.E.; Eitzinger, J.; Schaumberger, A.; Zahradnicek, P.; et al. Future agroclimatic conditions and implications for European grasslands. *Biol. Plant.* **2020**, *64*, 865–880. [CrossRef]
67. Wösten, J.; Pachepsky, Y.A.; Rawls, W. Pedotransfer functions: Bridging the gap between available basic soil data and missing soil hydraulic characteristics. *J. Hydrol.* **2001**, *251*, 123–150. [CrossRef]
68. Terrer, C.; Phillips, R.P.; Hungate, B.A.; Rosende, J.; Pett-Ridge, J.; Craig, M.E.; van Groenigen, K.J.; Keenan, T.F.; Sulman, B.N.; Stocker, B.D.; et al. A trade-off between plant and soil carbon storage under elevated CO_2. *Nature* **2021**, *591*, 599–603. [CrossRef]
69. Phukubye, K.; Mutema, M.; Buthelezi, N.; Muchaonyerwa, P.; Cerri, C.; Chalplot, V. On the impact of grassland management on soil carbon stocks: A worldwide meta-analysis. *Geoderma Reg.* **2022**, *28*, e00479. [CrossRef]
70. Tóth, G.; Jones, A.; Montanarella, L. The LUCAS topsoil database and derived information on the regional variability of cropland topsoil properties in the European Union. *Environ. Monit. Assess.* **2013**, *185*, 7409–7425. [CrossRef]
71. Poeplau, C.; Zopf, D.; Greiner, B.; Geerts, R.; Korvaar, H.; Thumm, U.; Don, A.; Heidkamp, A.; Flessa, H. Why does mineral fertilization increase soil carbon stocks in temperate grasslands? *Agric. Ecosyst. Environ.* **2018**, *265*, 144–155. [CrossRef]
72. Guillaume, T.; Bragazza, L.; Levasseur, C.; Libohova, Z.; Sinaj, S. Long-term soil organic carbon dynamics in temperate cropland-grassland systems. *Agric. Ecosyst. Environ.* **2021**, *305*, 107184. [CrossRef]
73. Kögel-Knabner, I.; Amelung, W. Soil organic matter in major pedogenic soil groups. *Geoderma* **2021**, *384*, 114785. [CrossRef]
74. Panagos, O.; Ballabio, C.; Himics, M.; Scarpa, S.; Matthews, F.; Bogonos, M.; Poesen, J.; Borrelli, P. Projections of soil loss by water erosion in Europe by 2050. *Environ. Sci. Policy* **2021**, *124*, 380–392. [CrossRef]
75. Zhao, Y.; Liu, Z.; Wu, J. Grassland ecosystem services: A systematic review of research advances and future directions. *Landsc. Ecol.* **2020**, *35*, 793–814. [CrossRef]

sustainability

MDPI

Review

Airport Runoff Water: State-of-the-Art and Future Perspectives

Anna Maria Sulej-Suchomska [1,*], Danuta Szumińska [2,*], Miguel de la Guardia [3], Piotr Przybyłowski [1] and Żaneta Polkowska [4]

[1] Department of Quality Management, Faculty of Management and Quality Sciences, Gdynia Maritime University, 81-225 Gdynia, Poland
[2] Department of Environmental Change and Geochemistry, Faculty of Geographical Sciences, Kazimierz Wielki University, 8 Kościelecki Sq., 85-033 Bydgoszcz, Poland
[3] Department of Analytical Chemistry, University of Valencia, 50th Dr. Moliner St., 46100 Burjassot, Spain
[4] Department of Analytical Chemistry, Faculty of Chemistry, Gdańsk University of Technology, 80-233 Gdańsk, Poland
* Correspondence: a.sulej-suchomska@wznj.umg.edu.pl (A.M.S.-S.); dszum@ukw.edu.pl (D.S.)

Abstract: The increase in the quantity and variety of contaminants generated during routine airport infrastructure maintenance operations leads to a wider range of pollutants entering soil and surface waters through runoff, causing soil erosion and groundwater pollution. A significant developmental challenge is ensuring that airport infrastructure meets high-quality environmental management standards. It is crucial to have effective tools for monitoring and managing the volume and quality of stormwater produced within airports and nearby coastal areas. It is necessary to develop methodologies for determining a wide range of contaminants in airport stormwater samples and assessing their toxicity to improve the accuracy of environmental status assessments. This manuscript aims to showcase the latest advancements (2010–2024 update) in developing methodologies, including green analytical techniques, for detecting a wide range of pollutants in airport runoff waters and directly assessing the toxicity levels of airport stormwater effluent. An integrated chemical and ecotoxicological approach to assessing environmental pollution in airport areas can lead to precise environmental risk assessments and well-informed management decisions for sustainable airport operations. Furthermore, this critical review highlights the latest innovations in remediation techniques and various strategies to minimize airport waste. It shifts the paradigm of soil and water pollution management towards nature-based solutions, aligning with the sustainable development goals of the 2030 Agenda.

Keywords: airport runoff water; water pollution; soil degradation; stormwater toxicity; best management practices (BMPs); sustainable development; environmental quality management

1. Introduction

Urban regions consist of a mix of impermeable and permeable surfaces, resulting in intricate patterns of rainfall and runoff over both space and time [1–3]. The prevalence of paved surfaces reduces water infiltration and increases runoff after precipitation events, potentially exerting a significant influence on streamflow patterns and flood risk [4–9]. As urbanization accelerates and climate change progresses, extreme weather events are becoming more frequent than in the past, leading to significant challenges for urban areas, such as increased instances of urban waterlogging [10]. Urbanization has changed land use patterns and the hydrological regime of urban areas, resulting in most rainfall becoming runoff [10,11]. Urban runoff, also known as urban stormwater, urban run-off, surface, or rainwater runoff encompasses the runoff, resulting primarily from precipitation on impermeable surfaces within urban environments, such as rooftops, pavements, roadways, and parking areas [12–14]. Furthermore, in regions experiencing seasonal snowfall, urban runoff also includes snowmelt [4–6]. This runoff carries large quantities of sediments and pollutants that have accumulated on urban surfaces into nearby water bodies, leading to the degradation of water quality when they mix with groundwater [5,7]. Urban runoff also exacerbates soil erosion, particularly given the often deteriorated state of urban soils [7–9,14,15].

A distinct type of urban stormwater is runoff from airports, which represents a very specific type of sample that is increasingly being subjected to analysis [16,17]. Recently, air transport has become one of the most important and fastest-growing sectors of the global economy. Although the industry was severely impacted by the COVID-19 pandemic, with dramatic declines in service rates and passenger numbers, there are clear signs that the industry is beginning to recover [18]. According to the International Air Transport Association (IATA), the total number of passengers in 2021 was 47% of the 2019 level. The numbers increased to 83% in 2022 and 94% in 2023, with further growth projected to reach 103% in 2024 and 111% in 2025 [18,19]. Despite the numerous benefits of the rapid growth in the air transport sector, all activities associated with airport operations result in environmental pollution [10,20]. The extensive use of hardened pavement in airport construction results in a high proportion of impervious surfaces, disrupting the natural hydrological cycle [10]. One of the major issues associated with airport operations is runoff water, which is generated when precipitation or atmospheric deposition washes over the surfaces of the airport apron during its use. Runoff water from airport areas can contain a wide spectrum of contaminants at varying concentration levels [21–27].

On an international scale, environmental protection, including the reduction of the negative impacts of airports on the environment, is a priority in sustainable development policy, in line with the goals and objectives of the 2030 Agenda for Sustainable Development [28]. The management of airport stormwater has emerged as a matter of national and international concern recently, with significant attention being paid to the effects of de-icing/anti-icing activities on the quality of runoff. Literature data emphasize that airport runoff should be regarded as a distinct pollutant stream, necessitating specialized treatment before being discharged into surface waters. To effectively address airport effluent quality within the framework of environmental safety, it is imperative to adopt a comprehensive and integrated management approach [29,30]. Efforts to mitigate the impacts of airport runoff water through the development and implementation of best management practices (BMPs) are hindered by an incomplete understanding of the composition of airport runoff water samples. The current literature indicates a relatively narrow range of research focused on the analysis of airport stormwater samples. These studies are limited to determining the content of basic summary parameters such as COD, BOD_5, total nitrogen, TOC [31–33], and organic compounds, especially from the groups of glycols, benzotriazoles, and PAHs [34–44]. The analysis of literature data leads to the conclusion that it is necessary to develop appropriate methodologies for determining a wide range of contaminants in airport stormwater samples. Furthermore, the fate, transport, and toxicity of pollutants in airport runoff are not yet fully understood [45]. There are clear gaps in the study results concerning the composition of airport runoff water samples and the whole effluent toxicity of runoff water from airports. The absence of standardized methods for measuring contaminants and evaluating toxicity in airport stormwater samples poses a significant barrier to the design of BMPs and generates uncertainty in the regulatory context [45].

The scientific community is currently concentrating on developing best management practices at airports to maximize the reduction of the negative impact associated with polluted runoff, particularly concerning water pollution, soil erosion, and soil degradation. This manuscript presents recent advancements in analytical methodologies, including green techniques, for determining a wide range of pollutants in airport runoff and assessing the overall toxicity of airport stormwater effluent. Such chemical-toxicological methodologies act as tools to identify the types and quantities of contaminants present in airport runoff. This leads to precise environmental risk assessments and well-informed management decisions for sustainable airport operations. Additionally, this review highlights the latest innovations in remediation techniques and various strategies to minimize airport waste, shifting the paradigm of soil and water pollution management towards nature-based solutions that promote the concept of Low-Impact Development (LID). Actions to protect the environment must be integrated and should encompass three dimensions: social, economic, and environmental. This approach will ensure the quality of life expected by society for current and future generations.

2. Basic Airport Stormwater Characterization

Urban runoff originates from various types of urban surfaces, each characterized by different levels of permeability and water retention capacity. The main categories of these surfaces include impervious surfaces (such as streets, roads, sidewalks, parking lots, and building roofs), semi-pervious surfaces (such as cobblestones, paving slabs, and permeable pavements), and pervious surfaces (such as lawns, green spaces, fences, and gravel paths). As previously mentioned, a specific type of stormwater runoff forms in urbanized areas, known as airport runoff. Airport runoff is generated from the diverse range of surfaces present within the airport grounds. Designated and properly prepared areas within an airport include surfaces where aircraft movement occurs, as well as zones intended for the parking of aircraft and technical service vehicles. These areas encompass runways, taxiways, and apron pavements, as well as pavements situated around buildings designated for terminals and aircraft parking. This type of surface includes hardstands, uncovered assembly areas, and roads designated for aircraft movement. Additionally, airport surfaces encompass those located within hangar and assembly halls. Airport surfaces also include natural surfaces, such as soil-based, grass, and turf surfaces. Based on the available data from various airports, Table 1 presents the averaged percentages of different types of airport surfaces along with their characteristics [26,30,46,47]. The proportional distributions may exhibit slight variations depending on the specific airport, its scale, the nature of operations conducted, and the configuration of its infrastructure.

Table 1. Percentage distribution of main airport operational areas along with surface characteristics.

Main Operational Areas of Airport Infrastructure	Proportion of the Airport's Total Operational Area [%]	Characteristics
(1) Runways	50–60	high-strength concrete/asphalt, durable for takeoffs and landings;
(2) Taxiways	20–30	concrete/asphalt, designed for smooth aircraft movement;
(3) Aircraft Parking Areas	15–25	reinforced concrete/asphalt, supports stationary aircraft during operations;
(4) Aprons	15–20	reinforced concrete, withstands heavy equipment use;
(5) Passenger Terminal Areas	10–20	asphalt/concrete, durable for high foot and vehicle traffic;
(6) Cargo Areas	10–15	durable concrete, supports cargo vehicles and equipment;
(7) Maintenance Areas	5–10	reinforced concrete, withstands heavy equipment use;
(8) De-icing zones	2–5	reinforced concrete, resistant to de-icing chemicals.

Airport pollutants emitted into the atmosphere undergo dispersion or transformation during atmospheric transport. However, most pollutants present in the air, soil, and water return to the Earth's surface through wet or dry deposition or by being absorbed by aerosol particles. This includes gaseous pollutants, which can be transferred through surface waters or soil [6,48–50]. A significant portion of pollutant transfer from the atmosphere to the Earth's surface occurs through wet deposition, specifically precipitation and atmospheric deposition [6,48,51]. Precipitation, atmospheric sediment and the resulting runoff flush the airport's surfaces during everyday operations [5,47,52]. Airports employ various chemical substances that may be introduced into stormwater pathways. This runoff, containing organic and inorganic contaminants, infiltrates the soil, surface waters, and groundwater, potentially affecting sources of drinking water [6,53,54]. Furthermore, during heavy precipitation events, this stormwater can lead to microbiological contamination of source water for public drinking supplies [55,56]. Intense rainfall leading to runoff water, particularly from areas near airports, has been associated with a high risk of acute gastrointestinal illness (AGI), notably in highly developed countries like France, the United Kingdom, Canada, the United States, and Australia [45,56].

The drainage design of the airport operational area is similar to that of urban areas, primarily utilizing buried pipes to discharge rainwater. The drainage system in the airfield area primarily includes covered ditches, open ditches, blind ditches, box culverts, V-ditches, and detention ponds [10]. Runoff water is conveyed away from airport areas through sewer systems to prevent localized flooding, as the majority of airports, regardless of traffic intensity and size, do not have their wastewater treatment plants [57]. Most airports do not even have their preliminary wastewater treatment facilities. Consequently, the generated stormwater, along with its contaminants, is directed into drainage ditches, significantly increasing the load and burden on the wastewater system [20,54,58,59]. In the absence of a treatment plant or if a treatment plant is functioning improperly, various contaminants may enter the air, soil, and surface waters along with runoff water [28,45,60,61].

3. Contaminants in Airport Runoff Waters

Various forms of anthropogenic pressure associated with the operation of any airport lead to the dispersal of pollutants in the environment where they undergo chemical, biochemical, and photochemical transformations [62,63]. Airport runoff water samples may contain a wide array of hazardous substances, including polycyclic aromatic hydrocarbons (PAHs), glycols, benzotriazoles (BTs or BTR), polychlorinated biphenyls (PCBs), per- and polyfluoroalkyl substances (PFAS), pesticides, biocides, detergents, potassium acetate, potassium formate, sodium formate, sodium acetate, phenols, formaldehyde, heavy metal ions, and cyanides [24–26]. The aforementioned pollutants come from de-icing/anti-icing operations, fuel spills, aircraft and installation cleaning, and, in short, all human activities related to passenger transport and the maintenance and operation of installations at airports. The main sources of pollution at airports are presented in Figure 1 [22–24,64–66].

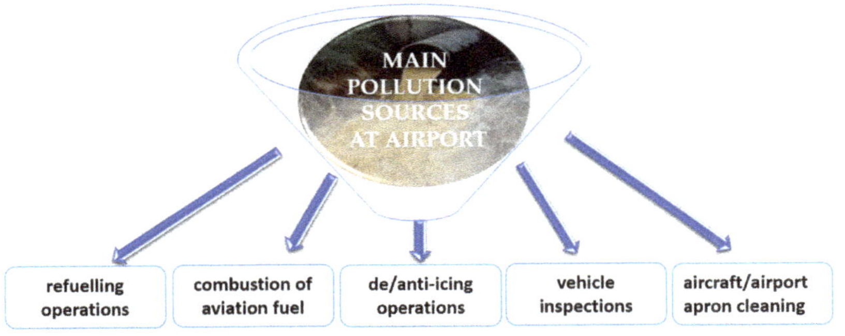

Figure 1. The most important sources of pollution at airports.

Our previous literature reviews on the analysis of xenobiotics in airport runoff water covered studies up to the year 2010 [24,25]. In this paper, an update is proposed, incorporating the latest data and research findings. Table 2 summarizes the studies published in the last years regarding the analysis of xenobiotics and determination of toxicity in airport runoff water samples. The target analytes, sampling site, along with the range of concentration and details about the bioassays and methodologies used for sample preparation and analyte determination, have been documented.

Based on data from global literature on the detection of xenobiotics in runoff water samples, it can be concluded that the most commonly analyzed inorganic analytes are cations, anions, total nitrogen, total phosphorus, and metals (Figure 2a). The most frequently detected organic compounds in runoff waters belong to the groups of PAHs, PCBs, PFCs, BTs, and glycols (Figure 2b). The data presented in Figure 2 indicate that different countries prioritize specific analytes in their airport runoff water testing. In Poland, the analysis of cations, anions, and total phosphorus among inorganic xenobiotics is a primary focus, while in Norway and the USA, there is a greater emphasis on the analysis of metals and total nitrogen. In terms of organic analytes, Poland leads in the detection of PAHs and PCBs, while the USA and China exhibit higher testing frequencies for glycols and PFCs.

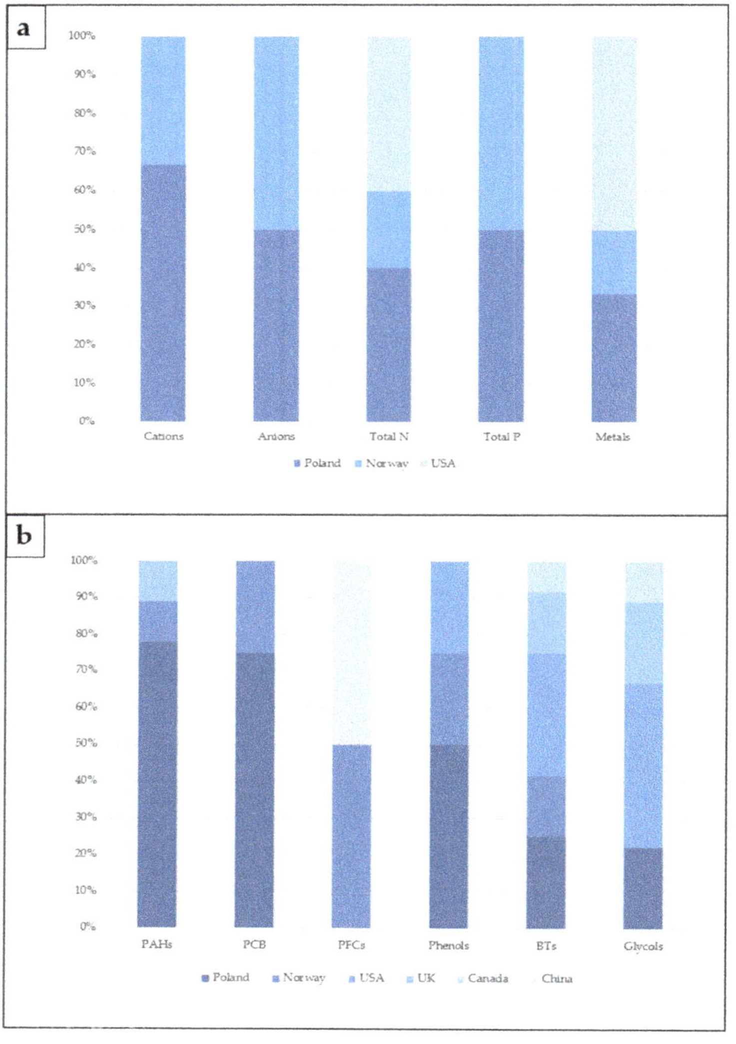

Figure 2. Percentage distribution of the most commonly analyzed inorganic (**a**) and organic (**b**) analytes in runoff water samples from airports across various countries.

The percentage distribution of toxicity tests conducted on airport runoff water samples across various countries is presented in Figure 3. In summary, based on the available literature, the Microtox® test and the Thamnotoxkit F™ are the most commonly used methods for assessing the toxicity of airport runoff water. The most extensive research on the toxicity of these water samples has been conducted in the USA and Poland.

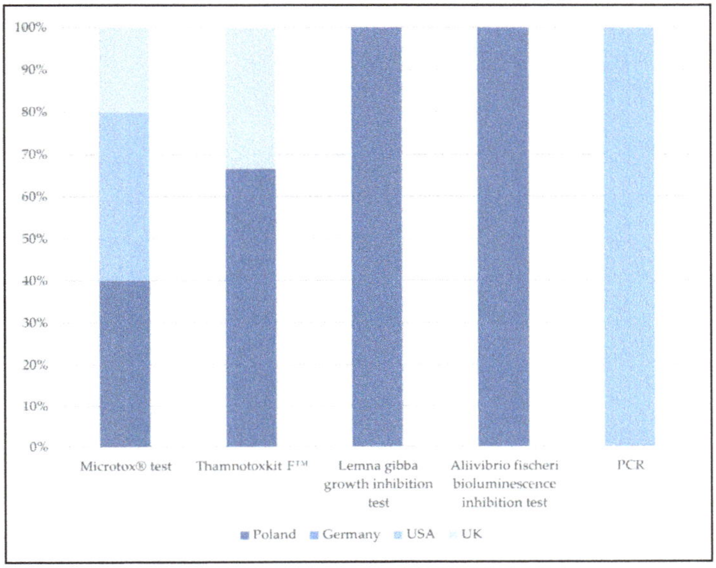

Figure 3. Percentage distribution of the conducted toxicity tests on airport runoff water samples in various countries.

National and international focus have intensified on the impact of de-icing/anti-icing activities on runoff quality [27,67,68]. De-icing/anti-icing of aircraft and airport platforms are fundamental operations to ensure safety at airports and during flights [69]. De-icing/anti-icing agents are routinely applied year-round at airports, across a wide range of temperatures [70]. Aircraft de-icing/anti-icing processes generally involve spraying the aircraft with de-icing and antifreeze fluids (ADAFs) that contain environmentally harmful and toxic chemical agents [65]. Chemical pollutants from the groups of glycols and benzotriazoles, primarily generated during de-icing/anti-icing operations on airport surfaces and aircraft, pose a significant threat to all elements of the environment due to their estrogenic potential, high toxicity, mutagenicity, and carcinogenicity [36,62,63,66,68,71–80]. Furthermore, recent studies indicate that exposure to benzotriazoles during pregnancy may elevate the risk of developing gestational diabetes mellitus (GDM) [62,81]. Due to their widespread use and poor degradability, these pollutants are commonly found in surface waters and wastewater treatment plant (WWTP) effluents globally.

Another persistent and unresolved issue is the generation of pollution associated with the combustion and spillage of aviation fuel. In this context, a group of compounds that require particular attention are polycyclic aromatic hydrocarbons. Compounds from the PAH group are primarily emitted during the combustion and spillage of aviation fuel, including fuel transport, refueling, and aircraft repairs [82]. Polycyclic aromatic hydrocarbons are recognized for their carcinogenic, teratogenic, and mutagenic properties, contributing to various health issues such as cancer, infertility, oxidative stress, and atherosclerosis [36,63,68,73–77,82]. PAHs present substantial health risks, and their elimination from the environment continues to be a worldwide priority [83].

Table 2. Recent literature data on the analysis of airport runoff water samples.

Sampling Site	Target Analytes/Summary Parameters	Analytical Methodology (Sample Preparation, Determination)/Bioassays	Range of Concentration	Reference
		XENOBIOTICS ANALYSIS		
Luleå Airport, Sweden	cations (Ca, Fe, K, Mg, Na, Al, Cu, As, Cd, Co, Cr, Mn, Ni, Pb, Zn), anions (Cl^-, SO_4^{2-}, NO^{3-}, NO^{2-});	-inductively coupled plasma atomic emission spectroscopy (ICP-AES); -inductively coupled plasma sector field mass spectrometry (ICP-SFMS);	<0.2–61.5 mg/L	[65]
	total phosphorus (total P)	colorimetric method	0.01–0.5 mg/L	
	total nitrogen (total N) NH_4^+-N, NO_3^-N, NO_2^--N	chemiluminescence method, nesslerization method, phenate method, Griess–Ilosvay method,	<MDL [2]–5.8 mg/L	
	benzene, toluene, ethylbenzene, and xylenes(BTEX), methyl tertiary butyl ether (MTBE), polycyclic aromatic hydrocarbons (PAH), polychlorinated biphenyls (PCB), aliphatics, chlorinated aliphatics, chlorinated pesticides, chlorinated benzenes, chlorophenols;	gas chromatography (GC), gas chromatography-mass spectrometry (GC-MS);	<MDL [2]–48 µg/L	
	perfluorinated compounds (PFCs) -perfluorooctane sulfonate (PFOS), -perfluorooctanoic acid (PFOA);	high-performance liquid chromatography (HPLC);	<0.01–0.23 µg/L	
International airport, Poland (high capacity) [1]	phenols; formaldehyde;	spectrophotometric method	0.052–163 mg/L	[17,52,84, 85]
	PCBs	solid phase extraction (SPE) coupled with gas chromatography-mass spectrometry (GC-MS);	0.350–116 µg/L	
	16 PAHs [naphthalene (Naph), acenaphthylene (Acy), acenaphthene (Ace), fluorene (Flu), phenanthrene (Ph), anthracene (An), fluoranthene (Flt), pyrene (Py), chrysene (Chry), benzo(b)fluoranthene (BbF), benzo(k)fluoranthene (BkF), benzo(a)pyrene (BaP), benzo(a)anthracene (BaA), indeno(1,2,3-cd)pyrene (InPy), dibenz(a,h)anthracene (DBahA), benzo(g,h,i)perylene (BghiP)];	-SPE-GC-MS; -headspace solid-phase microextraction (HS-SPME) coupled with comprehensive two-dimensional gas chromatography with time-of-flight mass spectrometry (GC × GC-TOF-MS);	2.25–187 µg/L	[52,86]
	benzotriazoles (BTs):1H-benzotriazole (1H-BT) 4-methylbenzotriazole (4-MeBT), 5-methylbenzotriazole (5-MeBT), 5,6-dimethyl-1H-benzotriazole (5,6-diMe-1H-BT),	SPE-GC–MS; HS-SPME-GC × GC-TOF-MS;	0.079–467,000 µg/L	[17,52]
	glycols: ethylene glycol (EG), propylene glycol (PG), diethylene glycol (DEG),	SPE-GC–MS, HS-SPME-GC–MS,	1.84–19,166 mg/L	[52,87]

Table 2. *Cont.*

Sampling Site	Target Analytes/Summary Parameters	Analytical Methodology (Sample Preparation, Determination)/Bioassays	Range of Concentration	Reference
		XENOBIOTICS ANALYSIS		
International airport, Poland (medium capacity) [1]	phenols, formaldehyde,	spectrophotometric method,	0.012–3.98 mg/L	[88]
	PCBs,	SPE-GC-MS,	0.325–1.39 µg/L	
	16 PAHs,	liquid–liquid extraction (LLE-GC–MS),	3.71–141 µg/L	
	EG, PG, DEG,	SPE-GC–MS, HS-SPME-GC–MS,	25.6–1184 mg/L	
	1H-BT, 4-MeBT, 5-MeBT, 5,6-diMe-1H-BT,	SPE-GC–MS, HS-SPME-GC × GC-TOF-MS,	0.402–156 µg/L	
International airport, Poland (low capacity) [1]	phenols, formaldehyde,	spectrophotometric method,	0.021–3.72 mg/L	[17,84]
	PCBs,	SPE-GC-MS,	0.424–1.71 mg/L	
	16 PAHs,	LLE-GC–MS, SPE-GC–MS, HS-SPME-GC × GC-TOF-MS,	2.3–190 µg/L	
	EG, PG, DEG,	SPE-GC–MS, HS-SPME-GC–MS,	1.9–19,166 mg/L	
	1H-BT, 4-MeBT, 5-MeBT, 5,6-diMe-1H-BT,	SPE-GC–MS, HS-SPME-GC × GC-TOF-MS,	0.079–467,000 µg/L	
International airport, United Kingdom (high capacity) [1]	16 PAHs,	SPE-GC–MS, HS-SPME-GC × GC-TOF-MS,	1.9–43 µg/L	[17,52]
	EG, PG, DEG,	SPE-GC–MS, HS-SPME-GC–MS,	3.97–270 mg/L	
	1H-BT, 4-MeBT, 5-MeBT, 5,6-diMe-1H-BT,	SPE-GC–MS, HS-SPME-GC × GC-TOF-MS,	0.083–7.67 µg/L	
Stuttgart airport, Germany	TOC	coulometric method,	300–1500 mg/L	[31]
Pearson Airport, Buttonville Airport, Canada	BTs: 1H-BT and its derivatives, UV stabilizers (BZT-UVs),	Solid phase extraction (SPE)-liquid chromatography-electrospray ionization-tandem mass spectrometry (HPLC-MS/MS)	8–3800 ng/L	[89]
Snohomish County Airport, Washington, USA	Zn (zinc),	-Inductively Coupled Plasma Optical Emission Spectroscopy (ICP-OES), -Scanning Electron Microscopy with an Energy Dispersive X-Ray (SEM & EDX)	1.25–25.0 µg/L	[90]
Białystok Airport, Poland	4-MeBT, 5-MeBT, 2-(3-tert-butyl-2-hydroxy-5-methylphenyl)−5-chloro-2H-benzotriazole (UV-326), 2-(2HBenzotriazol-2-yl)-4-(1,1,3,3-tetramethylbutyl)phenol (UV-329),	microextraction by ultrasound-assisted emulsification (USAEME)-GC–MS,	<MDL [2]–5.31 µg/L	[75]

Table 2. *Cont.*

Sampling Site	Target Analytes/Summary Parameters	Analytical Methodology (Sample Preparation, Determination)/Bioassays	Range of Concentration	Reference
		XENOBIOTICS ANALYSIS		
Białystok Airport, Poland	heavy metals (HMs): Cd, Pb, Ni, Cr, Cu, Zn,	filtration and mineralization, flame atomic absorption spectrometry (FAAS),	<1.00–213.6 µg/L	[91]
Beijing Capital International Airport (BC), Shanghai Pudong International Airport (SP) and Guangzhou Baiyun International Airport (GB), China	perfluoroalkyl and polyfluoroalkyl substances (PFAS),	SPE and ultra-high performance liquid chromatography coupled with triple quadrupole tandem mass spectrometry,	19.0–342 ng/L	[92]
General Mitchell International Airport, USA	1H-BT, 4-MeBT, 5-MeBT,	-SPE-GC-FID (before year 2007), -filtration (hydrophilic polytetrafluoroethylene (PTFE) syringe filters and HPLC-MS/MS (2007 and later),	<0.25–6600 µg/L	[68]
		TOXICOLOGICAL ANALYSIS		
International airport, Poland (high capacity) [1]	Microtox® test, Thamnotoxkit F™,		36.4% [3] 93.9% [4]	[93]
International airport, Poland (medium capacity) [1]	Microtox® test, Thamnotoxkit F™,		9.6% 79%	
International airport, Poland (low capacity) [1]	Microtox® test, Thamnotoxkit F™,		11.8% 97%	
International airport, United Kingdom(high capacity) [1]	Microtox® test, Thamnotoxkit F™,		NT [5] 20%	
Stuttgart airport, Germany	Lemna gibba growth inhibition test, Aliivibrio fischeri bioluminescence inhibition test,		31% 29%	[31]
Milwaukee Mitchell International Airport in Milwaukee, USA	Polymerase Chain Reaction (PCR): characterize relative *Sphaerotilus* abundance, determination of *sthA* sequence,		-	[94]

[1] Generic airport names are derived from cooperation agreements; [2] MDL—method detection limit; [3] The number of samples identified as toxic (relative to the total number of samples analyzed within the airport) using the Microtox® test; [4] The number of samples identified as toxic (relative to the total number of samples analyzed within the airport) using the Thamnotoxkit F™ test; [5] NT—non-toxic.

Runoff originating from airport surfaces can adversely affect the quality of soil and downstream water bodies, thereby disrupting the ecological integrity of aquatic, benthic, and terrestrial ecosystems [45,95]. A wide range of pollutants can accumulate in different components of the abiotic environment. Subsequently, a broad spectrum of pollutants can enter plants and, through the food chain, make their way into animal organisms and ultimately into human organisms. Chemicals emitted into the environment as a result of airport operations can cause numerous adverse effects immediately after exposure and later on, leading to delayed toxic response, mutagenic effects, and carcinogenic effects. The most effective way to reduce the environmental impact of pollutants emitted from airport operations is to prevent their generation [96]. For this reason, one of the primary tasks in airport infrastructure management is to continuously monitor and control the levels of pollutants emitted into the environment, along with measuring their toxicity [16,31,59,97–99].

4. Challenges in Studying Runoff Water from Airport Zones

In the field of pollutant analysis related to airport operations, new challenges continually arise. These challenges primarily involve the need to detect a wide range of analytes in samples with very complex and often variable matrix compositions. Special attention must be given to compounds from the groups of glycols, benzotriazoles, and PAHs, which pose a particular threat to all elements of the environment due to their toxicity and carcinogenicity.

Regardless of the location and scope of runoff water sample studies, conducting these analyses presents a substantial analytical challenge. One of the main challenges to tackle is the very low concentration levels of many contaminants in runoff water samples. Additionally, there is high variability in the levels of specific contaminants across different airports and at various times of the year. Furthermore, there is the possibility that the tested samples contain components with very similar physicochemical properties but significantly different toxicity towards the abiotic parts of the environment and the biota. Besides, the lack of standard techniques and equipment for sampling runoff water can significantly impact the reliability of measurement data. Another challenge is the difficulty in standardizing the measured results due to the geographical location of airports, varying intensity of air traffic, and changing meteorological conditions. There is also a lack of, or very limited access to, appropriate standard solutions. An important limitation is the shortage of reference materials with various metrological characteristics necessary for the calibration of control and measurement instruments, as well as the validation of analytical methods.

As previously mentioned, the literature indicates a relatively narrow scope of studies on airport stormwater samples [31–33]. A visualization of the number of research studies conducted on the determination of xenobiotics in airport stormwater around the world to date is shown in Figure 4. Although the first reports on the results of airport runoff water sample analyses, including our research, have appeared in the literature, this issue is still far from being fully understood and recognized [31,68]. However, it can be stated with full confidence that such material objects are gaining increasing interest as a source of information about the potential negative impact of airport activities on environmental conditions [14,17,68,100].

The literature review (Table 2) highlights the need for developing methodologies to detect a wide range of contaminants in airport stormwater and assess their toxicity, driven by ecotoxicological concerns and the goal of improving environmental assessments [30]. These methodologies must have such metrological parameters that enable the detection, identification, and quantification of a wide range of xenobiotics present in the samples at trace and even ultra-trace levels. This approach not only involves typical chemical methods and speciation analysis but also focuses on determining the toxic, mutagenic, and carcinogenic properties of collected samples. Such an integrated chemical-toxicological approach to assessing environmental pollution in airport areas allows for a comprehensive understanding of environmental quality [93,101–103]. In the context of the sustainable development agenda, it is crucial to have a comprehensive knowledge base and the necessary tools to effectively address a range of environmental and social issues.

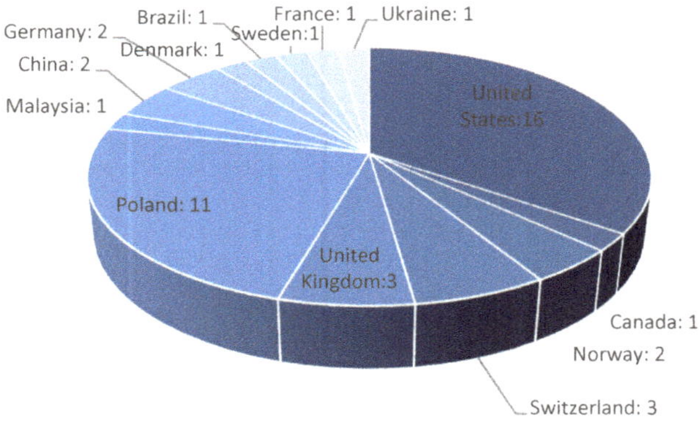

Figure 4. Pie chart of scientific publications on xenobiotic determination in airport runoff water samples, categorized by country and number of publications.

5. Integrated Chemical and Ecotoxicological Approach to Assessing Environmental Pollution in Airport Areas

The increase in the quantity and diversity of pollutants generated during daily airport infrastructure maintenance operations contributes to a broad spectrum of contaminants entering soil and surface waters through runoff. A significant developmental challenge is to ensure that airport infrastructure meets high-quality standards. In this regard, it is essential to have appropriate tools for monitoring and controlling the quantity and quality of water generated within airports and surrounding coastal areas. To determine the extent of environmental contamination resulting from airport operations, identify the emitted pollutants, and assess their toxicity levels, an integrated approach to the analysis of runoff water samples is necessary. The integrated (chemical and ecotoxicological) approach to assessing environmental pollution in airport areas will provide a detailed understanding of environmental quality. This will improve the management of environmental factors and mitigate the impact of various pollutants on the environment.

5.1. New Determination Methodologies for Monitoring the Airport Runoff Waters

A comprehensive analysis must be considered to obtain detailed qualitative and quantitative information about the pollutants in airport stormwater samples. Additionally, it is crucial to track the environmental fate of these pollutants. This includes monitoring the processes of transport, and the chemical, photochemical, and biological transformations of various compound groups that enter runoff waters due to specific activities and operations associated with airport functioning. Considering the available literature, the problems, and challenges associated with runoff water analysis indicate that, so far, there are a limited number of procedures for monitoring and controlling the quality of this new type of environmental sample. Additionally, these procedures typically have limited applications and are often time- and labor-intensive.

To obtain reliable information about the analyte content in airport stormwater samples with a complex matrix composition, high toxicity to living organisms, and many interfering compounds, it is essential to develop appropriate isolation and enrichment techniques. Additionally, the choice of the proper determination method is also crucial, with chromatography techniques playing an increasingly important role in this field. The use of different extraction and determination techniques for detecting xenobiotics in airport runoff water can lead to significantly different outcomes. A comparative analysis of the efficiency of various analytical protocols can be conducted in studies of these material objects. An additional significant aspect is that, optimizing the extraction and determination process conditions. This helps to effectively remove interfering components, enhance detectability, facilitate the separation of analytes during chromatographic analysis, and reduce the operational time of control and measurement equipment during the sample extract testing phase [104].

Based on a review of the available literature, it can be concluded that procedures primarily based on liquid-liquid extraction (LLE) and solid-phase extraction (SPE) have been successfully applied to determine xenobiotics in airport water. Standard analytical techniques used to detect and quantify xenobiotics in airport runoff water samples include gas chromatography-mass spectrometry (GC-MS), gas chromatography-flame ionization detection (GC-FID), and high-performance liquid chromatography (HPLC). These techniques, often employed after isolation or derivatization steps, also encompass high-performance liquid chromatography-mass spectrometry (HPLC-MS), high-performance liquid chromatography with ultraviolet detection (HPLC-UV), and high-performance liquid chromatography-tandem mass spectrometry (HPLC-MS/MS). Additionally, other methods, such as gel permeation chromatography (GPC), thin-layer chromatography (TLC), atomic absorption spectroscopy (AAS), and inductively coupled plasma mass spectrometry (ICP-MS), are also employed [24,25].

The growing interest of researchers in developing and optimizing modern, reliable, green, and powerful analytical procedures for the determination of xenobiotics in airport runoff water led to intensive research in this field (Table 2). This manuscript provides an update on green sample treatment methodologies applied recently for the extraction of xenobiotics from airport runoff water samples. Solid-phase microextraction (SPME) is a modern alternative to traditional extraction methods that have been used for isolating analytes from runoff water samples. SPME is a simple, convenient, and solvent-free extraction technique that combines extraction, concentration, and sample introduction into a single step. The experimental data demonstrate that the HS-SPME technique is an effective method for extracting de-icing/anti-icing compounds (glycols) from airport stormwater samples without the need for analyte derivatization. The newly developed methodology, based on SPME and GC–MS, described in this study facilitated the sensitive, accurate, precise, relatively rapid, and cost-effective identification and quantification of propylene glycol, ethylene glycol, and diethylene glycol in airport stormwater samples [87]. Another example is the development of a reliable and accurate analytical procedure based on HS-SPME coupled with comprehensive two-dimensional gas chromatography with time-of-flight mass spectrometry (GC × GC-TOF-MS) for the simultaneous determination of polycyclic aromatic hydrocarbons in airport stormwater samples. In this case, it is particularly important to highlight the usefulness of the GC × GC-TOF-MS technique for determining analytes in the extracts. Gas chromatography is a commonly used technique for separating and identifying organic compounds. Despite the high resolution offered by capillary columns, GC is often inadequate for separating the wide spectrum of interfering substances commonly found in environmental samples. The peak capacity of GC × GC is significantly higher compared to one-dimensional GC [105,106]. The use of comprehensive two-dimensional gas chromatography greatly enhances analyte separation. Furthermore, it is feasible to separate analytes from interfering components in stormwater, including, for example, petroleum-derived substances, polycyclic aromatic hydrocarbons, and polychlorinated biphenyls. Another example of a newly developed modern analytical procedure for determining analytes in stormwater runoff is the HS-SPME-GC × GC-TOF-MS-based method for identifying anti-corrosive compounds, including 1H-BT, 4-MeBT, 5-MeBT, and 5,6-diMe-1H-BT, in airport stormwater from various European airports [17]. Yet another innovation in the effective determination of analytes in airport runoff water samples is the use of a microextraction by ultrasound-assisted emulsification, combined with gas chromatography–mass spectrometry. This is used for the determination of low-molecule benzotriazoles such as 1H-BT, 4Me-BT, 5Me-BT, 5Cl-BT, and benzotriazole UV stabilizers (UV-326, UV-329) [78].

Despite some progress in green analytical methods for detecting xenobiotics in airport stormwater, the issue remains insufficiently recognized, necessitating ongoing development and enhancement of existing procedures. It is essential to conduct extensive, continuous monitoring and systematic control of the impact of runoff water on various environmental components, and indirectly on living organisms. Airports urgently require robust and economical solutions to tackle the challenges posed by the pollution of stormwater [26].

5.2. Assessment of the Toxic Potential of Airport Runoff Water

Biomonitoring of runoff water samples from airports must also be considered, as it allows for the analysis of the impact of pollutants on metabolic and physiological processes, mortality and survival rates, development, reproduction, and bioaccumulation capacity (resulting from bioconcentration and biomagnification) in living organisms. Additionally, it facilitates the observation of changes in flora and fauna in the monitored airport areas. Biomonitoring of runoff water samples from airports can be performed using various biotests, including toxicity tests (e.g., *Pseudokirchneriella subcapitata*, *Daphnia magna*), embryotoxicity tests (e.g., *Danio rerio*), enzymatic tests (e.g., EROD, *Vibrio fischeri*), the comet assay for cytogenetic analysis, carcinogenicity tests (e.g., SOS chromotest), and mutagenicity tests (e.g., Salmonella/Ames test).

Based on the analysis of available literature, it can be concluded that there are significant gaps in the research devoted to estimating the overall environmental impact of airport runoff water. Most studies report on the potential toxicological effects of aircraft de-icing/anti-icing solutions on various aquatic organisms, including *Pimephales promelas*, *Daphnia magna*, *Daphnia pulex*, *Ceriodaphnia dubia*, *Photobacterium phosphoreum*, *Lemna gibba*, and *Aliivibrio fischeri*. The results of the conducted research are mostly limited to the analysis of a single airport, where sample collection campaigns were not systematically organized. Globally, only a limited number of studies have been conducted to estimate the toxicity of airport runoff water (Table 2). To summarize, studies conducted to estimate the overall environmental impact of runoff water around airports involved assessing the toxic potential of runoff waters from five European international airports with varying capacities. Procedures based on the use of *Aliivibrio fischeri*, *Lemna gibba*, *Vibrio fischeri*, and *Thamnocephalus platyurus* bioassays were employed as tools for monitoring and controlling the quality of airport runoff water [31,84,93]. The results of the conducted studies indicate that runoff water samples from the monitored airports were often characterized by high or very high toxicity, according to the toxicity classification by G. Persoone et al. [107]. Recommended management actions were presented for all monitored locations at the airports. Based on the data obtained, it was determined that immediate corrective actions are necessary to improve the effectiveness and efficiency of environmental management systems, particularly in aircraft de-icing zones, airport aprons, runways, passenger terminal areas, and parking zones [93].

Furthermore, adopting an additional method to evaluate the environmental risk of xenobiotics in airport stormwater samples can help identify and understand the sources, pathways, and impacts of these environmental stressors. Accordingly, results from lab-scale toxicological studies are incorporated into this process [9,78]. However, there is a lack of comprehensive knowledge regarding the toxicological data of various xenobiotics. Based on the available literature, it can be concluded that there is a very limited scope of research focused on estimating the environmental risk associated with airport stormwater runoff. Attempts were made to estimate the environmental risk of low molecular weight benzotriazoles and benzotriazole UV stabilizers exclusively [17,78]. This approach can contribute to the development of effective strategies to prevent or mitigate the impacts of pollutants on soil erosion and degradation, as well as on surface water and groundwater resources.

Determining the toxicity levels of airport runoff waters and estimating the environmental risk of different pollutants generated within airports' activities is crucial for effective management. However, the available literature does not provide the data required for a comparative analysis of the toxicity of runoff water generated at airports of different sizes in different geographical regions and with varying levels of throughput. This area of research is not yet well understood and requires further analysis and numerous research tasks to be conducted [31,93].

6. Sustainable Stormwater Management at Airports

Numerous stakeholders have expressed their commitment to mitigate the risks that the airport's stormwater contamination may pose to the environment with potential economic, social, health, and public safety consequences [69,99]. Considering the increasing pressure to mitigate adverse effects and conserve soil and water resources, airports must manage their activities and operations to reduce water consumption and limit pollutants emitted in runoff during routine maintenance activities. Airports are responsible for safeguarding soil, surface water, and groundwater resources [108]. In light of reports presented by the U.S. Federal Aviation Administration (FAA), the International Civil Aviation Organization (ICAO), and the European Union Aviation Safety Agency (EASA), managing the quantity and quality of runoff water poses a particularly challenging task for airports [96,109,110]. It is important that effectively manage stormwater in a manner that does not compromise aircraft safety while simultaneously adhering to a multitude of federal, state, and local regulations aimed at safeguarding soil and water resources [111].

Stormwater Best Management Practices are measures, techniques, or structural controls designed to manage the quantity and improve the quality of stormwater in a cost-effective manner. To ensure compliance with relevant water management regulations, BMPs at or near airports must be carefully designed, considering site-specific physical conditions, watershed dimensions, runoff dynamics, peak flow rates, and targeted water quality objectives [112–114]. A key aspect in designing BMPs is the consideration of the characteristics of the underlying surface composition across the airport's various functional areas. Figure 5 presents a hypothetical airport layout with the main operational areas of airport infrastructure highlighted, as numbered in Table 1, which are typically found at airports. Considering the characteristics of the underlying surface composition across the airport's various functional areas in BMP planning can significantly contribute to preventing, reducing, and controlling erosion, sediment, and pollutant runoff from different operational zones of the airport.

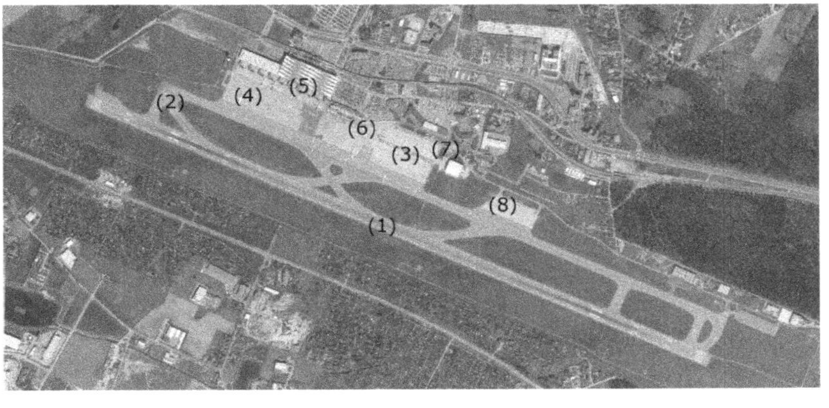

Figure 5. Hypothetical airport layout illustrating the main operational areas of airport infrastructure, numbered (1)–(8) according to the descriptions in Table 1.

Developed progressively since the late 1990s, the concept of Low-Impact Development focuses on managing stormwater and pollution through decentralized, small-scale source control measures (evapotranspiration, detention, retention, infiltration, drainage, and exfiltration of runoff water) [115]. The goal is to mimic the natural hydrological cycle as closely as possible within the development area [10]. LID can manage runoff, control soil erosion, and improve water quality. Selecting appropriate LIDs and evaluating their potential environmental impacts before implementation is critically important. Hydrology and soil erosion models have been extensively used to predict how hydrology and water quality respond to BMPs, including LIDs, in urban systems [115]. The traditional approach to airport rainwater management focused on rapid discharge through pipelines, canals, and pumps. Nowadays, airports are adopting strategies such as expanding green spaces, constructing large-capacity water storage facilities, installing green roofs, and prioritizing the treatment and reuse of rainwater resources [10,116,117]. Table 3 summarizes recent studies on technological solutions for treating and recycling airport runoff waters, as well as innovative approaches to ensure the environmental sustainability of airport development. Our previous literature review on remediation technologies, including forced hot air de-icing, the appropriate use of less glycol, recycling, and infrared de-icing, covered studies up to the year 2010 [24]. This literature review proposes an update that includes the latest data and newly developed solutions for managing airport runoff.

There are numerous techniques available for removing contaminants from stormwater, which can be categorized as gravity or coalescence separators, electrical separators, mechanical separators, and chemical separators [118,119]. Membranes have been considered a technology that can address this issue and mitigate global water shortages [117]. Another viable solution is the corrugated plate oil–water separation technology, which integrates gravity and coalescence separation methods. This technology has garnered significant attention from researchers and companies due to its feasibility, low operating costs, and high efficiency. When it comes to oil separators, Corrugated Plate Interceptors (CPI) are the most practical and cost-effective option, given their low initial and operating costs, as well as their ability to perform effectively under a wide range of operating conditions. In contrast, electrical and mechanical separators require additional power and chemical separation techniques demand a subsequent process to remove the chemicals used [119]. Numerous studies indicate that a drainage scheme integrating LID with pumping stations provides the most effective runoff management [115].

Various strategies have been proposed and implemented at the local, regional, and global levels with the aim of mitigating the negative effects of airport stormwater on the environment [109,110]. The summarized technological solutions for treating and recycling airport runoff waters, as well as the proposed innovative approaches, can enhance the environmental safety of airport operations. These proposals form the basis for developing strategies to ensure the environmental sustainability of airport development. However, there is still insufficient regulatory scope regarding the necessity of determining and removing a wide spectrum of toxic xenobiotics from runoff water generated during airport activities. In some cases, the implementation and enforcement of existing regulations and management actions are lacking. Poor international cooperation and insufficient participation of states in regional initiatives are also contributing factors [96]. The primary concern is the inclusion of compounds from the benzotriazole group, along with various other xenobiotics generated at airports, in the lists of substances particularly hazardous to the aquatic environment. These should be addressed in the regulations of the European Parliament and Council (EC), the Food and Agriculture Organization (FAO), and the International Civil Aviation Organization (ICAO).

A global data repository should be established to document the degree and type of pollution in water runoff from airports. This repository will enable comprehensive digital data analysis and management for existing and planned airports worldwide. Additionally, such a database will support the concept of Low-Impact Development and facilitate improved modeling and management of environmental quality. An exemplary instance of utilizing such a data source is the World Council on City Data (WCCD) database. This database has facilitated the creation of valuable and insightful studies on the quality of life in cities across Europe and globally, based on sustainable development indicators as defined by the ISO 37120 standard [120,121].

Table 3. Techniques and solutions for managing airport runoff.

Remediation/ Pollution Minimization Techniques/Solutions	Operating Principle	Materials/Equipment	Airport	Benefit	References
Vertical flow planted sand beds (VFPB)	organic load removal from glycol contaminated runoff water,	-addition of optimal quantity of nitrogen, phosphorus;	Orly Airport, France;	removal of high suspended matter content, high hydrocarbon loads;	[122]
Polyvinylidene fluoride membranes with tin (IV) dioxide (SnO_2) additives	water filtration using membrane technology with polyvinylidene fluoride (PVDF) polymer mixed with Tin (IV) Dioxide (SnO_2),	-Scanning Electron Microscopy (SEM); -Clean Water Permeability (CWP) tool;	Husein Sastranegara Airport, Indonesia;	increasing concentrations of PVDF and SnO_2;	[117]
Electrochemical treatment	electrolysis (3–4 V);	-Ti/PbO_2 electrode; -Electron microcopy; -XRD analysis;	model runoff water containing ethylene glycol;	electrochemical oxidation of ethylene glycol, reduces COD values;	[123]
Additional treatment (Coalescenting filter –Electrolyser deaerator –Sorption filter)	-coalescence-removes water insoluble pollutants; -cavitation; -electrochemical treatment (electrolyzer-deaerator and electrohydrocyclones); -photocatalytic ozonation; -sorption;	-coalescenting filter; -hypochlorite unit; -cavitation unit; -electrochemical treatment (electrolyzer-deaerator, electrohydrocyclones); -ozonator; -UV unit; -sorption filter;	International Airport Kharkiv, Ukraine;	-reduces: weighted substances, BOD, oil products, COD, phosphorus (total), iron; -nitrogen (total);	[20]
Corrugated Plate Interceptor (CPI)	-separating the oils from water;	-oil Water Separator; -inlet, outlet pipes; -sludge collectionsumps;	Malaysian airport;	the system separates and safely retains oils from water until removal;	[119,124]
A wastewater recycling centre	water recycling with a regular monitoring program for quality at the airport;		Osaka's Kansai International Airport, Japan;	increases the annual recycled water ratio (66.1%); -monitoring of COD, total nitrogen, total phosphates.	[108]

7. Conclusions and Future Trends

Given the increase in the amount and variety of pollutants produced during routine airport infrastructure maintenance operations, it is crucial to focus on the contaminants entering soil and surface waters through runoff. These pollutants contribute to soil erosion and significant water pollution. A major developmental challenge is ensuring that airport infrastructure complies with high-quality environmental management standards. In this context, it is essential to have suitable tools for monitoring and managing the volume and quality of wastewater produced within airports and adjacent coastal areas. The analysis of pollutants from airport operations presents new challenges, primarily involving the detection of diverse analytes in complex, variable matrices with a high level of accuracy, sensitivity, and selectivity.

An integrated chemical and ecotoxicological approach to assessing environmental pollution in airport areas enables the monitoring and control of airport runoff water quality, as well as the creation and updating of databases for international environmental information systems. This approach allows for rapid ecological risk assessment associated with airport wastewater streams and estimates the long-term effects of organism exposure to toxic substances present in airport runoff waters.

In sustainable stormwater management, various strategies have been implemented to mitigate the negative effects of airport stormwater on the environment. However, the regulatory framework remains insufficient regarding the need to detect or remove a wide spectrum of toxic xenobiotics from runoff water generated by airport activities. It is necessary to establish permissible levels for the presence of the aforementioned chemical compounds in water runoff from airports due to their toxicological impact on living organisms and the environment. The primary concern is the inclusion of compounds from the benzotriazole, glycol, and PAH groups, along with various other toxic xenobiotics generated at airports, in the lists of substances particularly hazardous to soil and aquatic environments. Furthermore, these substances should be subjected to rigorous monitoring and control within the context of operational airports and their surrounding areas. Comprehensive monitoring studies should consider the research schedule in relation to airports in different geographical regions with varying meteorological conditions, intensities of air traffic, and seasons for sampling campaigns and subsequent analysis. The article's content addresses the principles of Low-Impact Development and explores the quality of life paradigm, understood as the comprehensive interaction of environmental, social, and economic factors.

Author Contributions: Conceptualization, methodology, software, validation, formal analysis, investigation, data curation, writing—original draft preparation, editing, visualization, project administration, review and editing, A.M.S.-S.; review and editing, supervision, D.S.; review and editing, supervision, M.d.l.G.; review and editing, supervision, P.P.; conceptualization, review and editing, supervision, Ż.P. All authors have read and agreed to the published version of the manuscript.

Funding: Founded by the Ministry of Science under the "Regional Initiative of Excellence Program", agreement No. RID/SP/0023/2024/01. Additionally, the research was funded by Gdynia Maritime University under grant no. KZJ-WZNJ/2024/PZ/04.

Institutional Review Board Statement: Not applicable.

Informed Consent Statement: Not applicable.

Data Availability Statement: Data are contained within the article.

Conflicts of Interest: The authors declare no conflicts of interest.

References

1. Essien, A.E.; Guo, Y.; Khafagy, M.; Dickson-Anderson, S.E. Design and Hydrologic Performance Estimation of Highway Filter Drains Using a Novel Analytical Probabilistic Model. *Sci. Rep.* **2024**, *14*, 2350. [CrossRef]
2. Barszcz, M.P. Application of Dynamic and Conceptual Models for Simulating Flow Hydrographs in an Urbanized Catchment under Conditions of Controlled Outflow from Stormwater Tanks. *J. Water Clim. Chang.* **2021**, *12*, 3899–3914. [CrossRef]
3. Sahoo, S.N.; Sreeja, P. Development of Flood Inundation Maps and Quantification of Flood Risk in an Urban Catchment of Brahmaputra River. *ASCE-ASME J. Risk Uncertain. Eng. Syst. Part A Civ. Eng.* **2017**, *3*, A4015001. [CrossRef]
4. Ferreira, C.S.; Kalantari, Z.; Seifollahi-Aghmiuni, S.; Ghajarnia, N.; Rahmati, O.; Solomun, M.K. Chapter 21—Rainfall-Runoff-Erosion Processes in Urban Areas. In *Precipitation*; Rodrigo-Comino, E., Ed.; Elsevier: Amsterdam, The Netherlands, 2021; pp. 481–498, ISBN 9780128226995.
5. Müller, A.; Österlund, H.; Marsalek, J.; Viklander, M. The Pollution Conveyed by Urban Runoff: A Review of Sources. *Sci. Total Environ.* **2020**, *709*, 136125. [CrossRef]
6. Marsalek, J. *Urban Water Cycle Processes and Interactions*; CRC Press: Boca Raton, FL, USA, 2014; ISBN 9781482288544.
7. Pistocchi, A. A Preliminary Pan-European Assessment of Pollution Loads from Urban Runoff. *Environ. Res.* **2020**, *182*, 109129. [CrossRef]
8. Irvine, K.N.; Chua, L.H.C.; Hua'an, Z.; Qi, L.E.; Xuan, L.Y. Nature-Based Solutions to Manage Particle-Bound Metals in Urban Stormwater Runoff: Current Design Practices and Knowledge Gaps. *J. Soils Sediments* **2023**, *23*, 3671–3688. [CrossRef]
9. Perrodin, Y.; Boillot, C.; Angerville, R.; Donguy, G.; Emmanual, E. Ecological Risk Assessment of Urban and Industrial Systems: A Review. *Sci. Total Environ.* **2011**, *409*, 5162–5176. [CrossRef]
10. Peng, J.; Yu, L.; Zhong, X.; Dong, T. Study on Runoff Control Effect of Different Drainage Schemes in Sponge Airport. *Water Resour. Manag.* **2022**, *36*, 1043–1055. [CrossRef]
11. Cristiano, E.; Deidda, R.; Viola, F. EHSMu: A New Ecohydrological Streamflow Model to Estimate Runoff in Urban Areas. *Water Resour. Manag.* **2020**, *34*, 4865–4879. [CrossRef]
12. Walsh, C.J.; Booth, D.B.; Burns, M.J.; Fletcher, T.D.; Hale, R.L.; Hoang, L.N.; Livingston, G.; Rippy, M.A.; Roy, A.H.; Scoggins, M.; et al. Principles for Urban Stormwater Management to Protect Stream Ecosystems. *Freshw. Sci.* **2016**, *35*, 398–411. [CrossRef]
13. Walsh, C.J.; Fletcher, T.D.; Burns, M.J. Urban Stormwater Runoff: A New Class of Environmental Flow Problem. *PLoS ONE* **2012**, *7*, e45814. [CrossRef] [PubMed]
14. Zhang, Y.; Zhao, W.; Chen, X.; Jun, C.; Hao, J.; Tang, X.; Zhai, J. Assessment on the Effectiveness of Urban Stormwater Management. *Water* **2021**, *13*, 4. [CrossRef]
15. Jia, Y.; Bakken, L.R.; Breedveld, G.D.; Aagaard, P.; Frostegård, Å. Organic Compounds That Reach Subsoil May Threaten Groundwater Quality; Effect of Benzotriazole on Degradation Kinetics and Microbial Community Composition. *Soil Biol. Biochem.* **2006**, *38*, 2543–2556. [CrossRef]

16. Marthinussen, N. Modelling Airport Runoff Containing De-icing Chemicals Case Study: Stavanger Airport Sola. Master's Thesis, Norwegian University of Science and Technology, Faculty of Engineering Department of Civil and Environmental Engineering, Trondheim, Norway, 2019; pp. 1–42.
17. Sulej-Suchomska, A.M.; Koziol, K.; Polkowska, Ż. Comprehensive Analysis and Environmental Risk Assessment of Benzotriazoles in Airport Stormwater: A HS-SPME-GC × GC-TOF-MS-Based Procedure as a Tool for Sustainable Airport Runoff Water Management. *Sustainability* **2024**, *16*, 5152. [CrossRef]
18. Abdi, Y.; Li, X.; Càmara-Turull, X. Firm Value in the Airline Industry: Perspectives on the Impact of Sustainability and COVID-19. *Humanit. Soc. Sci. Commun.* **2023**, *10*, 294. [CrossRef] [PubMed]
19. The International Air Transport Association IATA. Available online: https://www.iata.org/en/pressroom/2022-releases/2022-03-01-01/ (accessed on 30 August 2024).
20. Zhelnovach, G.; Belokon, K.; Barabash, O.; Dychko, A. Airport Runoff Management: Engineering Solutions. *Ecol. Eng. Environ. Technol.* **2022**, *23*, 230–240. [CrossRef]
21. Fan, H.; Tarun, P.K.; Chen, V.C.P.; Shih, D.T.; Rosenberger, J.M.; Kim, S.B.; Horton, R.A. Data-Driven Optimization for Dallas Fort Worth International Airport Deicing Activities. *Ann. Oper. Res.* **2018**, *263*, 361–384. [CrossRef]
22. Switzenbaum, M.S.; Veltman, S.; Mericas, D.; Wagoner, B. Theodore Schoenberg Best Management Practices for Airport Deicing Stormwater. *Chemosphere* **2001**, *43*, 1051–1062. [CrossRef]
23. Koryak, M.; Stafford, L.J.; Reilly, R.J.; Hoskin, R.H.; Haberman, M.H. The Impact of Airport Deicing Runoff on Water Quality and Aquatic Life in a Pennsylvania Stream. *J. Freshw. Ecol.* **1998**, *13*, 287–298. [CrossRef]
24. Sulej, A.M.; Polkowska, Z.; Namieśnik, J. Pollutants in Airport Runoff Waters. *Crit. Rev. Environ. Sci. Technol.* **2012**, *42*, 1691–1734. [CrossRef]
25. Sulej, A.M.A.M.; Polkowska, Z.; Namieśnik, J. Analysis of Airport Runoff Waters. *Crit. Rev. Anal. Chem.* **2011**, *41*, 190–213. [CrossRef]
26. Shi, X.; Quilty, S.M.; Long, T.; Jayakaran, A.; Fay, L.; Xu, G. Managing Airport Stormwater Containing Deicers: Challenges and Opportunities. *Front. Struct. Civ. Eng.* **2017**, *11*, 35–46. [CrossRef]
27. Mielcarek, A.; Rodziewicz, J.; Janczukowicz, W.; Ostrowska, K. The Kinetics of Pollutant Removal through Biofiltration from Stormwater Containing Airport De-Icing Agents. *Appl. Sci.* **2021**, *11*, 1724. [CrossRef]
28. United Nations. *Transforming Our World: The 2030 Agenda for Sustainable Development*; United Nations: New York, NY, USA, 2015.
29. Craig, G.R.; van Arkel, G.; Williams, J.B.; LeRoux, J. How to Develop Chemical Criteria for Airport Stormwater. *J. Water Manag. Model.* **1998**, *6062*, 309. [CrossRef]
30. Luther, L. *Environmental Impacts of Airport Operations, Maintence, and Expansion*; Division, R.S.I., Ed.; Congressional Research Service: Washington, DC, USA, 2007; Volume 298.
31. Calvo, O.C.; Quaglia, G.; Mohiley, A.; Cesarini, M.; Fangmeier, A. Assessing Potential Aquatic Toxicity of Airport Runoff Using Physicochemical Parameters and Lemna Gibba and Aliivibrio Fischeri Bioassays. *Environ. Sci. Pollut. Res.* **2020**, *27*, 40604–40617. [CrossRef]
32. Siedlecka, E.; Downar, D. Jakość Wód z Rejonu Portu Lotniczego Gdańsk- Trójmiasto. *Chem. i Inżynieria Ekol.* **2004**, *11*, 557–567.
33. Krzemieniowski, M.; Białowiec, A.; Zieliński, M. The Effectiveness of the Storm Water Treatment Plant at Warsaw Frederick Chopin Airport. *Arch. Environ. Prot.* **2006**, *32*, 25–33.
34. *EPA-821-R-00-016*; USEPA Preliminary Data Summary. USEPA-United States Environmental Protection Agency: Washington, DC, USA, 2000.
35. Mohiley, A.; Franzaring, J.; Calvo, O.C.; Fangmeier, A. Potential Toxic Effects of Aircraft De-Icers and Wastewater Samples Containing These Compounds. *Environ. Sci. Pollut. Res.* **2015**, *22*, 13094–13101. [CrossRef]
36. Han, X.; Xie, Z.; Tian, Y.; Yan, W.; Miao, L.; Zhang, L.; Zhu, X.; Xu, W. Spatial and Seasonal Variations of Organic Corrosion Inhibitors in the Pearl River, South China: Contributions of Sewage Discharge and Urban Rainfall Runoff. *Environ. Pollut.* **2020**, *262*, 114321. [CrossRef]
37. Corsi, S.R.; Geis, S.W.; Bowman, G.; Failey, G.; Rutter, T. Di Aquatic Toxicity of Airfield-Pavement Deicer Materials and Implications for Airport Runoff. *Environ. Sci. Technol.* **2009**, *43*, 40–46. [CrossRef]
38. Corsi, S.R.; Zitomer, D.J.; Field, J.; Cancilla, D.A. Nonylphenol Ethoxylates and Other Additives in Aircraft Deicers, Antiicers, and Waters Receiving Airport Runoff. *Environ. Sci. Technol.* **2003**, *37*, 4031–4037. [CrossRef]
39. Corsi, S.R.; Harwell, G.R.; Geis, S.W.; Bergman, D. Impacts of Aircraft Deicer and Anti-Icer Runoff on Receiving Waters from Dallas/Fort Worth Intrnational Airport, Texas, USA. *Environ. Toxicol. Chem.* **2006**, *25*, 2890–2900. [CrossRef] [PubMed]
40. Corsi, S.R.; Hall, D.W.; Geis, S.W. Aircraft and Runway Deicers at General Mitchell International Airport, Milwaukee, Wisconsin, USA. 2. Toxicity of Aircraft and Runway Deicers. *Environ. Toxicol. Chem.* **2001**, *20*, 1483–1490. [CrossRef] [PubMed]
41. Breedveld, G.D.; Roseth, R.; Sparrevik, M.; Hartnik, T.; Hem, L.J. Persistence of the De-Icing Additive Benzotriazole at an Abandoned Airport. *Water Air Soil Pollut.* **2003**, *3*, 91–101. [CrossRef]
42. Zitomer, D.H. *Waste Aircraft Deicing Fluid: Managemen and Conversion to Methane*; Marquette University: Milwaukee, WI, USA, 2001.
43. Cancilla, D.A.; Baird, J.C.; Rosa, R. Detection of Aircraft Deicing Additives in Groundwater and Soil Samples from Fairchild Air Force Base, a Small to Moderate User of Deicing Fluids. *Bull. Environ. Contam. Toxicol.* **2003**, *70*, 868–875. [CrossRef]

44. Mahvi, A.; Mardani, G. Determination of Phenanthrene in Urban Runoff of Tehran, Capital of Iran. *J. Environ. Health Sci. Eng.* **2005**, *2*, 5–11.
45. Grant, S.; Rekhi, N.V.; Pise, N.R.; Reeves, R.L. *A Review of the Contaminants and Toxicity Associated with Particles in Stormwater Runoff*; University of California, Ed.; University of California: Sacramento, CA, USA, 2015.
46. European Union Aviation Safety Agency; the European Environment Agency (EEA); EUROCONTROL. *European Aviation Environmental Report 2019*; EASA: Cologne, Germany, 2019; pp. 1–108, ISBN 978-92-9210-214-2.
47. Barash, S.; Covington, J.; Tamulonis, C. *Preliminary Data Summary Airport Deicing Operations (Revised)*; United States Environmental Protection Agency: Washington, DC, USA, 2000; Volume 4303.
48. Polkowska, Ż. *Atmospheric Deposits. Problems and Chalanges*; Cieśliński, J., Ed.; Gdańsk University of Technology Publishing: Gdansk, Poland, 2008; ISBN 0416-7341.
49. Nunes, L.M.; Zhu, Y.G.; Stigter, T.Y.; Monteiro, J.P.; Teixeira, M.R. Environmental Impacts on Soil and Groundwater at Airports: Origin, Contaminants of Concern and Environmental Risks. *J. Environ. Monit.* **2011**, *13*, 3026–3039. [CrossRef]
50. Polkowska, Z.; Górecki, T.; Namieśnik, J. Determination of Atmospheric Pollutants in Wet Deposition. *Environ. Rev.* **2011**, *19*, 185–213. [CrossRef]
51. Guo, K.; Wang, L. On the Water Environment System in Water Sensitive Area—Building of Sponge Airport Stormwater System in Beijing New Airport. *IOP Conf. Ser. Earth Environ. Sci.* **2018**, *146*, 012029. [CrossRef]
52. Sulej, A.M.A.M.; Polkowska, Z.; Astel, A.; Namieśnik, J. Analytical Procedures for the Determination of Fuel Combustion Products, Anti-Corossive Compounds, and de-Icing Compounds in Airpot Runoff Water Samples. *Talanta* **2013**, *117*, 158–167. [CrossRef]
53. Riley, K.; Cook, R.; Carr, E.; Manning, B. A Systematic Review of the Impact of Commercial Aircraft Activity on Air Quality near Airports. *City Environ. Interact.* **2021**, *11*, 100066. [CrossRef] [PubMed]
54. Walters, E.; Spuller, K. Effects of Airport Runoff Pollution on Water Quality in Bay Area Sites near San Francisco and Oakland Airports. *J. Emerg. Investig.* **2021**, *4*, 8–11. [CrossRef] [PubMed]
55. Nguyen, M.A.; Norström, K.; Wiberg, K.; Gustavsson, J.; Josefsson, S.; Ahrens, L. Seasonal Trends of Per- and Polyfluoroalkyl Substances in River Water Affected by Fire Training Sites and Wastewater Treatment Plants. *Chemosphere* **2022**, *308*, 136467. [CrossRef] [PubMed]
56. De Roos, A.J.; Kondo, M.C.; Robinson, L.F.; Rai, A.; Ryan, M.; Haas, C.N.; Lojo, J.; Fagliano, J.A. Heavy Precipitation, Drinking Water Source, and Acute Gastrointestinal Illness in Philadelphia, 2015–2017. *PLoS ONE* **2020**, *15*, 2015–2017. [CrossRef] [PubMed]
57. Council of Ministers. *State Environmental Policy 2030—Development Strategy in the Area of Environment and Water Management, September 6*; Monitor Polski, Official Journal of the Republic of Poland: Warsaw, Poland, 2019.
58. Rodziewicz, J.; Mielcarek, A.; Janczukowicz, W.; Bryszewski, K.; Ostrowska, K. Treatment of Wastewater Containing Runway De-Icing Agents in Biofilters as a Part of Airport Environment Management System. *Sustainability* **2020**, *12*, 3608. [CrossRef]
59. Baxter, G.; Srisaeng, P.; Wild, G. An Assessment of Airport Sustainability: Part 3-Water Management at Copenhagen Airport. *Resources* **2019**, *8*, 135. [CrossRef]
60. Voutsa, D.; Hartmann, P.; Schaffner, C.; Giger, W. Benzotriazoles, Alkylphenols and Bisphenol A in Municipal Wastewaters and in the Glatt River, Switzerland. *Environ. Sci. Pollut. Res.* **2006**, *13*, 333–341. [CrossRef]
61. Lapointe, M.; Rochman, C.M.; Tufenkji, N. Sustainable Strategies to Treat Urban Runoff Needed. *Nat. Sustain.* **2022**, *5*, 366–369. [CrossRef]
62. Shi, Z.Q.; Liu, Y.S.; Xiong, Q.; Cai, W.W.; Ying, G.G.; Shi, Z.-Q.; Liu, Y.-S.; Xiong, Q.; Cai, W.-W.; Ying, G.-G. Occurrence, Toxicity and Transformation of Six Typical Benzotriazoles in the Environment: A Review. *Sci. Total Environ.* **2019**, *661*, 407–421. [CrossRef]
63. Minella, M.; De Laurentiis, E.; Pellegrino, F.; Prozzi, M.; Bello, F.D.; Maurino, V.; Minero, C. Photocatalytic Transformations of 1h-Benzotriazole and Benzotriazole Derivates. *Nanomaterials* **2020**, *10*, 1835. [CrossRef]
64. Chávez-mejía, A.C.; Magaña-lópez, R.; Durán-álvarez, J.C.; Jiménez-cisneros, B.E. Indexed by CAMES—Conseil Africain et Malgache pour l'Enseignement. *Int. J. Environ. Agric. Biotechnol. IJEAB* **2019**, 16–32. [CrossRef]
65. Jia, Y.; Ehlert, L.; Wahlskog, C.; Lundberg, A.; Maurice, C. Water Quality of Stormwater Generated from an Airport in a Cold Climate, Function of an Infiltration Pond, and Sampling Strategy with Limited Resources. *Environ. Monit. Assess.* **2018**, *190*, 1–18. [CrossRef] [PubMed]
66. Cantwell, M.G.; Sullivan, J.C.; Burgess, R.M. *Benzotriazoles: History, Environmental Distribution, and Potential Ecological Effects*; Elsevier: Amsterdam, The Netherlands, 2015; Volume 67, ISBN 9780444632999.
67. Jaiyeola, A.T. The Management and Treatment of Airport Rainwater in a Water-Scarce Environment. *Int. J. Environ. Sci. Technol.* **2017**, *14*, 421–434. [CrossRef]
68. Olds, H.T.; Corsi, S.R.; Rutter, T.D. Benzotriazole Concentrations in Airport Runoff Are Reduced Following Changes in Airport Deicer Formulations. *Integr. Environ. Assess. Manag.* **2022**, *18*, 245–257. [CrossRef]
69. Surya, B.; Hamsina, H.; Ridwan, R.; Baharuddin, B.; Menne, F.; Fitriyah, A.T.; Rasyidi, E.S. The Complexity of Space Utilization and Environmental Pollution Control in the Main Corridor of Makassar City, South Sulawesi, Indonesia. *Sustainability* **2020**, *12*, 9244. [CrossRef]
70. Konieczka, R. The airport de-icing of aircrafts. *Sci. J. Silesian Univ. Technol.-Ser. Transp.* **2015**, *86*, 55–64.
71. O'Donnell, M. *Management of Airport Industrial Waste*; U.S. Department of Transportation, Federal Aviation Administration, Ed.; Federal Aviation Administration: Washington, DA, USA, 2008.

72. Gigger, W.; Schaffner, C.; Kohler, H.E. Benzotriazole and Tylotriazole as Aguatic Contaminants. 1. Imput and Occurrence in Rivers and Lakes. *Environ. Sci. Technol* **2006**, *40*, 7186–7192. [CrossRef]
73. Pillard, D.; Dufresne, D.; Hernandez, M. Toxicity of Benzotriazole and Benzotriazole Derivatives to Three Aquatic Species. *Water Res.* **2001**, *35*, 557–560. [CrossRef]
74. Hart, D.D.S.; Davis, L.C.L.; Erickson, L.L.E.; Callender, T.T.M. Sorption and Partitioning Parameters of Benzotriazole Compounds. *Microchem. J.* **2004**, *77*, 9–17. [CrossRef]
75. Kotowska, U.; Sokołowska, J.S.; Piekutin, J. Simultaneous Determination of Low Molecule Benzotriazoles and Benzotriazole UV Stabilizers in Wastewater by Ultrasound—Assisted Emulsification Microextraction Followed by GC—MS Detection. *Sci. Rep.* **2021**, *11*, 10098. [CrossRef]
76. Corsi, S.R.; Geis, S.W.W.; Rice, C.P.; Sheesley, R.J.; Loyo-rosales, J.E.; Rice, C.P.; Sheesley, R.J.; Failey, G.G.; Cancilla, D.A. Characterization of Aircraft Deicer and Anti-Icer Components and Toxicity in Airport Snowbanks and Snowmelt Runoff. *Environ. Sci. Technol.* **2006**, *40*, 3195–3202. [CrossRef] [PubMed]
77. Rahim, A.A.; Saad, B.; Osman, H.; Yahya, S.; Talib, K.M. Simultaneous Determination of Diethylene Glycol, Diethylene Glycol Monoethyl Ether, Coumarin and Caffeine in Food Items by Gas Chromatography. *Food Chem.* **2011**, *126*, 1412–1416. [CrossRef]
78. Struk-Sokołowska, J.; Gwoździej-Mazur, J.; Jurczyk, Ł.; Jadwiszczak, P.; Kotowska, U.; Piekutin, J.; Canales, F.A.; Kaźmierczak, B. Environmental Risk Assessment of Low Molecule Benzotriazoles in Urban Road Rainwaters in Poland. *Sci. Total Environ.* **2022**, *839*, 156246. [CrossRef] [PubMed]
79. Krautsieder, A.; Sharifi, N.; Madden, D.C.; Sonke, J.; Routh, A.F.; Clarke, S.M. Corrosion Inhibitor Distribution on Abrasive-Blasted Steels. *J. Colloid Interface Sci.* **2023**, *634*, 336–345. [CrossRef]
80. Alvey, J.K.; Hagedorn, B.; Dotson, A. Benzotriazole Enrichment in Snowmelt Discharge Emanating from Engineered Snow Storage Facilities. *Water Environ. Res.* **2016**, *88*, 510–520. [CrossRef]
81. Zhou, Y.; Qu, J.; Liu, W.; Liao, J.; Li, Y.; Zhao, H.; Li, J. Early Pregnancy Exposure to Benzotriazoles and Benzothiazoles in Relation to Gestational Diabetes Mellitus: A Prospective Cohort Study. *Environ. Int.* **2020**, *135*, 105360. [CrossRef]
82. Wicke, D.; Matzinger, A.; Sonnenberg, H.; Caradot, N.; Schubert, R.L.; Dick, R.; Heinzmann, B.; Dünnbier, U.; von Seggern, D.; Rouault, P. Micropollutants in Urban Stormwater Runoff of Different Land Uses. *Water* **2021**, *13*, 1312. [CrossRef]
83. Janarthanam, V.A. Hazards of Polycyclic Aromatic Hydrocarbons: A Review on Occurrence, Detection, and Role of Green Nanomaterials on the Removal of PAH from the Water Environment. *Environ. Monit. Assess.* **2023**, *195*, 1531. [CrossRef]
84. Sulej, A.M.; Polkowska, Ż.; Wolska, L.; Cieszynska, M.; Namieśnik, J. Toxicity and Chemical Analyses of Airport Runoffwaters in Poland. *Environ. Sci. Process. Impacts* **2014**, *16*, 1083–1093. [CrossRef]
85. Sulej, A.; Polkowska, Z.; Namieśnik, J. Contaminants in Airport Runoff Water in the Vicinities of Two International Airports in Poland. *Pol. J Environ. Stud.* **2012**, *21*, 725–739.
86. Sulej-Suchomska, A.M.A.M.; Polkowska, Z.; Chmiel, T.; Dymerski, T.M.; Kokot, Z.J.Z.J.; Namieśnik, J. Solid Phase Microextraction-Comprehensive Two-Dimensional Gas Chromatography-Time-of-Flight Mass Spectrometry: A New Tool for Determining PAHs in Airport Runoff Water Samples. *Anal. Methods* **2016**, *8*, 4509–4520. [CrossRef]
87. Sulej-Suchomska, A.M.; Polkowska, Z.; Kokot, Z.J.; de la Guardia, M.; Namieśnik, J. Determination of Antifreeze Substances in the Airport Runoff Waters by Solid-Phase Microextraction and Gas Chromatography-Mass Spectrometry Method. *Microchem. J.* **2016**, *126*, 466–473. [CrossRef]
88. Sulej-Suchomska, A.M.; Polkowska, Ż.; Przyjazny, A.; Kokot, Z.J.; Namieśnik, J.; Polkowska, Ż.; Namieśnik, J. Determination of Fuel Combustionproduct in Airport Runoff Water Samples Using Liquid–Liquid Extraction with Gas Chromatography–Spectrometry *Int. J. Environ. Sci. Technol.* **2016**, *13*, 1475–1488. [CrossRef]
89. Parajulee, A.; Lei, Y.D.; Desilva, A.O.; Cao, X.; Mitchell, C.P.J.; Wania, F. Assessing the Source-to-Stream Transport of Benzotriazoles during Rainfall and Snowmelt in Urban and Agricultural Watersheds. *Environ. Sci. Technol.* **2017**, *51*, 4191–4198. [CrossRef]
90. Long, T.; Zou, L. Extraction of Zinc from Airport Stormwater Runoff Using Oyster Shells. *Coll. Aviat. Rev.* **2019**, *37*, 45–58. [CrossRef]
91. Bełcik, M.; Grzegorzek, M.; Canales, F.A.; Struk-Sokołowska, J.; Kaźmierczak, B. Examination of Interactions between Heavy Metals and Benzotriazoles in Rainwater Runoff and Snowmelt in an Urban Catchment in Poland. *Water Resour. Ind.* **2024**, *31*, 100236. [CrossRef]
92. Liu, J.; Zhao, Z.; Li, J.; Hua, X.; Zhang, B.; Tang, C.; An, X.; Lin, T.L. Emerging and Legacy Perfluoroalkyl and Polyfluoroalkyl Substances (PFAS) in Surface Water around Three International Airports in China. *Chemosphere* **2023**, *344*, 140360. [CrossRef] [PubMed]
93. Sulej-Suchomska, A.M.; Przybyłowski, P.; Polkowska, Ż. Potential Toxic Effects of Airport Runoff Water Samples on the Environment. *Sustainability* **2021**, *13*, 7490. [CrossRef]
94. Nott, M.A.; Driscoll, H.E.; Takeda, M.; Vangala, M.; Corsi, S.R.; Tighe, S.W. Advanced Biofilm Analysis in Streams Receiving Organic Deicer Runoff. *PLoS ONE* **2020**, *15*, e0227567. [CrossRef]
95. Ellis, B. Urban Runoff Quality in the UK: Problems, Prospects and Procedures. *Appl. Geogr.* **1991**, *11*, 187–200. [CrossRef]
96. Kale, U.; Jankovics, I.; Nagy, A.; Rohács, D. Towards Sustainability in Air Traffic Management. *Sustainability* **2021**, *13*, 5451. [CrossRef]
97. Pavliukh, L. Perspectives of Wastewater Treatment By Microalgae At an Airport. *Sci. Technol.* **2021**, *50*, 147–152. [CrossRef]

98. Velautham, K.D. Storm Water Management in Airport Using Oil Water Separator System. *Turkish J. Comput. Math. Educ.* **2021**, *12*, 1014–1020. [CrossRef]
99. Brtnický, M.; Pecina, V.; Baltazár, T.; Galiová, M.V.; Baláková, L.; Beś, A.; Radziemska, M. Environmental Impact Assessment of Potentially Toxic Elements in Soils near the Runway at the International Airport in Central Europe. *Sustainability* **2020**, *12*, 7224. [CrossRef]
100. Paige, T.; De Silva, T.; Buddhadasa, S.; Prasad, S.; Nugegoda, D.; Pettigrove, V. Background Concentrations and Spatial Distribution of PFAS in Surface Waters and Sediments of the Greater Melbourne Area, Australia. *Chemosphere* **2024**, *349*, 140791. [CrossRef]
101. Ferguson, L.; Corsi, S.R.; Geis, S.W.; Anderson, G.; Joback, K.; Gold, H.; Mericas, D.; Cancilla, D.A. *Formulations for Aircraft and AirfieldDeicing and Anti-Icing: AquaticToxicity and Biochemical OxygenDemand*; University of South Carolina: Columbia, SC, USA, 2008.
102. Pillard, D.A. Comparative Toxicity of Formulated Glycol Deicers and Pure Ethylene and Propylene Glycol to Ceriodaphnia Dubia and Pimephales Promela. *Environ. Toxicol. Chem.* **1995**, *14*, 311–315. [CrossRef]
103. Pillard, D.; DuFrense, D. Toxicilty of Formulated Glycol Deicers and Ethylene and Propylene Glycol to Lactuca Sativa, Lolium Perenne, Selenastrum Capricornutum, and Lemma Minor. *Arch. Environ. Contam. Toxicol.* **1999**, *37*, 29–35. [CrossRef]
104. Esteve-Turrillas, F.A.; Garrigues, S.; de la Guardia, M. Green Extraction Techniques in Green Analytical Chemistry: A 2019–2023 up-Date. *TrAC Trends Anal. Chem.* **2024**, *170*, 117464. [CrossRef]
105. Vaye, O.; Ngumbu, R.S.; Xia, D. A Review of the Application of Comprehensive Two-Dimensional Gas Chromatography MS-Based Techniques for the Analysis of Persistent Organic Pollutants and Ultra-Trace Level of Organic Pollutants in Environmental Samples. *Rev. Anal. Chem.* **2022**, *41*, 63–73. [CrossRef]
106. Stefanuto, P.-H.; Smolinska, A.; Focant, J.-F. Advanced Chemometric and Data Handling Tools for GC×GC-TOF-MS: Application of Chemometrics and Related Advanced Data Handling in Chemical Separations. *Trends Anal. Chem.* **2021**, *139*, 116251. [CrossRef]
107. Persoone, G.; Marsalek, B.; Blinova, I.; Törökne, A.; Zarina, D.; Manusadzianas, L.; Nalecz-Jawecki, G.; Tofan, L.; Stepanova, N.; Tothova, L.; et al. A Practical and User-friendly Toxicity Classification System with Microbiotests for Natural Waters and Wastewaters. *Environ. Toxicol.* **2003**, *18*, 395–402. [CrossRef] [PubMed]
108. Baxter, G.; Srisaeng, P.; Wild, G. An Assessment of Sustainable Airport Water Management: The Case of Osaka's Kansai International Airport. *Infrastructures* **2018**, *3*, 54. [CrossRef]
109. EPA-Environmental Protection Agency. EPA Part 122—EPA Administered Permit Programs: The National Pollutant Discharge Elimination System. 33 U.S.C. 1251. Available online: https://www.govinfo.gov/link/uscode/33/1251 (accessed on 30 August 2024).
110. ICAO. *Water Management at Airports*; ICAO: Montreal, Canada, 2019.
111. Xiong, C.; Beckmann, V.; Tan, R. Effects of Infrastructure on Land Use and Land Cover Change (LUCC): The Case of Hangzhou International Airport, China. *Sustainability* **2018**, *10*, 2013. [CrossRef]
112. Ayivi, F.; Jha, M.K. Estimation of Water Balance and Water Yield in the Reedy Fork-Buffalo Creek Watershed in North Carolina Using SWAT. *Int. Soil Water Conserv. Res.* **2018**, *6*, 203–213. [CrossRef]
113. Paraschi, E.P.; Georgopoulos, A.; Kaldis, P. Airport Business Excellence Model: A Holistic Performance Management System. *Tour. Manag.* **2019**, *72*, 352–372. [CrossRef]
114. Aldrees, A.; Dan, S. Application of Analytical Probabilistic Models in Urban Runoff Control Systems ' Planning and Design: A Review. *Water* **2023**, *15*, 1640. [CrossRef]
115. Guo, T.; Srivastava, A.; Flanagan, D.C.; Liu, Y.; Engel, B.A.; Mcintosh, M.M. Evaluation of Costs and Efficiencies of Urban Low Impact Development (LID) Practices on Stormwater Runoff and Soil Erosion in an Urban Watershed Using the Water Erosion Prediction Project (WEPP) Model. *Water* **2021**, *13*, 2076. [CrossRef]
116. Chikatamarla RK, C.S. Doubling the Capacity of Kempegowda International Airport in Bengaluru India. *Proc. Inst. Civ. Eng. Civ. Eng.* **2021**, *174*, 37–44. [CrossRef]
117. Fazal, M.R.; Mataram, A. Polyvinylidene Fluoride Membranes With Tin (Iv) Dioxide (Sno2) Additives: Enhancing Water Treatment for Airport Eco Green. *J. Airpt. Eng. Technol.* **2023**, *3*, 68–74. [CrossRef]
118. Han, Y.; He, L.; Luo, X.; Lü, Y.; Shi, K.; Chen, J.; Huang, X. A Review of the Recent Advances in Design of Corrugated Plate Packs Applied for Oil–Water Separation. *J. Ind. Eng. Chem.* **2017**, *53*, 37–50. [CrossRef]
119. Velautham, K.D.; Chelliapan, S.; Kamaruddin, S.A.; Meyers, J.L. Design of Oil Water Separator for the Removal of Hydrocarbon from Stormwater Contaminated with Jet-Fuel. *J. Adv. Res. Fluid Mech. Therm. Sci.* **2022**, *92*, 162–176. [CrossRef]
120. Przybyłowski, P.; Przybyłowski, A.; Kałaska, A. Utility Method as an Instrument of the Quality of Life Assessment Using the Examples of Selected European Cities. *Energies* **2021**, *14*, 2770. [CrossRef]
121. Przybyłowski, A.; Kałaska, A.; Przybyłowski, P. Quest for a Tool Measuring Urban Quality of Life: ISO 37120 Standard Sustainable Development Indicators. *Energies* **2022**, *15*, 2841. [CrossRef]
122. Branchu, P.; Gres, L.; Mougin, F.; Le Blanc, M.; Lucas, E.; Mars, B. French Airport Runoff Pollution Management (Water and Sludge): Toward a New Approach Based on Constructed Wetlands? Case of Aéroports de Paris—Orly (France). *Water Pract. Technol.* **2014**, *9*, 20–32. [CrossRef]

123. Magomedova, D.S.; Alimirzayeva, Z.M.; Magomedova, A.G.; Isaev, A.B.; Kharlamova, T.A. Electrochemical Treatment of Airport Runoff Water Containing Ethylene Glycol. *Chem. Probl.* **2022**, *20*, 109–115. [CrossRef]
124. Velautham, K.D.; Chelliapan, S.; Kamaruddin, S.A.; Meyers, J.L. Design Requirements for the Treatment of Stormwater Contaminated with Jet Fuel Oil Using Corrugated Plate Interceptor. *Egypt. J. Chem.* **2022**, *65*, 1–10. [CrossRef]

sustainability

MDPI

Article

Temporal Dynamics of Ecosystem Service Value in Qinghai Province (2000–2018): Trends, Drivers, and Implications

Lan Li [1], Yilu Zhao [2,*], Xuan Li [1], Wankui Ni [2,3], Kuanyao Zhao [1,*], Ziguang He [1], Mohamed EL-Sayed Abuarab [4], Ahmed Elbeltagi [5] and Adnan Ahmed [6]

[1] College of Civil Engineering and Architecture, Huanghuai University, Zhumadian 463000, China; lilan@huanghuai.edu.cn (L.L.)
[2] College of Geological Engineering and Geomatics, Chang'an University, Xi'an 710054, China
[3] Key Laboratory of Western China's Mineral Resources and Geological Engineering, Ministry of Education, Chang'an University, Xi'an 710054, China
[4] Faculty of Agriculture, Cairo University, Giza 12613, Egypt
[5] Agricultural Engineering Department, Faculty of Agriculture, Mansoura University, Mansoura 35516, Egypt
[6] College of Water and Environment, Chang'an University, Xi'an 710054, China
* Correspondence: zhaoyilu1999@163.com (Y.Z.); ky.zhao@huanghuai.edu.cn (K.Z.)

Abstract: Ecosystem service value (ESV) represents the benefits that people obtain directly or indirectly from ecosystems. Based on the land use/land cover (LULC) data of Qinghai Province in 2000, 2010, and 2018, combined with the equivalent factor method, land use dynamic degree, land use transition matrix, and coefficient of sensitivity (CS), the ESV of Qinghai Province is calculated. This study found that ESV in Qinghai Province exhibited significant variations. It has increased from CNY 750.77 billion to CNY 772.4 billion from 2000 to 2018. The results show that grassland and unused land are the main land use types, with a total coverage of more than 88% in the study area. The land use dynamic degree result demonstrates that the dynamic degree of built-up land is the highest. The CS analysis found that the CSs of the ESV of each land use type in Qinghai Province were less than 1 during 3 periods, 2000–2010, 2010–2018, and 2000–2018, indicating that the ESV in the area lacks elasticity and the result is reliable. This study can provide scientific support for the sustainable development of ecosystems and land use planning in Qinghai Province under global climate change.

Keywords: LUCC; equivalent factor method; ecosystem service valuation; elasticity; Qinghai province

1. Introduction

Qinghai Province, located in the west of China and the northeast of the Qinghai-Tibet Plateau (QTP), is also the headstream of many rivers in China, such as the Yellow River, the Yangtze River, the Lancang River, etc. These rivers represent important water sources in Asia, a strategic base for China's water resources security, and an ecological barrier for environmental security and regional sustainable development [1]. Qinghai Province boasts rich ecosystems and diverse land use types, serving as an important component of the ecological security barrier. Despite this, it has been facing challenges such as ecological degradation and biodiversity loss in recent years, making it an ideal region for studying changes in ecosystem service value and their influencing factors. Unique geographic and climatic characteristics have a profound influence on a series of ecosystem services, such as climate regulation, water supply, and the carbon cycle in China and even globally [2,3]. Meantime, Qinghai Province is also the area with the most fragile ecosystem. Regional development and utilization of natural resources resulted in grassland degradation, soil erosion, land desertification, and other problems [4,5]. At present, the research related to the ESV of Qinghai Province is not comprehensive and mainly focuses on particular regions or specific ecosystems [6]. Overall, in consideration of the protection and utilization of resources and the rapid socioeconomic development [7], Qinghai Province has accumulated rich experience in ecological protection, which holds significant reference value for other similar regions. Qinghai Province proposed the implementation of an "ecological province" strategy to promote the coordinated development of the ecological environment and economy. Therefore, an in-depth analysis of land use/land cover change (LUCC) and its impact on ESV can provide a scientific basis for the sustainable use of land resources in this region to realize the strategic goal of an "ecological province".

ES refers to the benefits that the Natural Capital provides for human well-being [8,9], which has a significant value in maintaining ecological balance and coping with environmental pollution, climate change, and biodiversity conservation [9]. Ecosystem services can be divided into four categories: supporting services, regulating services, provisioning services, and cultural services; among these, the supporting services are at the base of the other three services [10]. The combined effect of LUCC and climate is an important driving force of biodiversity reduction [11,12]. As the carrier of the natural ecosystem, if LUCC develops in a negative direction, it may not just reduce the regulation ability of the ecosystem, but also affect land productivity and cause a series of adverse consequences [13–15]. The development of the economy and the acceleration of urbanization and industrialization [16,17] influenced a series of environmental problems, such as pollution, waste of resources, sharp decrease in biodiversity, frequent occurrence of extreme weather, etc. [18]. Consequently, the ecosystem gradually loses its balance, and its services begin to deteriorate. The relationship between LUCC and ES is becoming closer, with the proven impact of LUCC on ES [19]. Therefore, the quantitative evaluation of ES based on long-term LUCC is helpful in guiding human activities to prevent land degradation and curb the loss of biodiversity [20–22].

Currently, the evaluation of ESV has become the frontier and hotspot of international geography, ecology, and related disciplines [23,24]. It is important in the construction of ecological barriers, division of ecological protection red lines, and coordination of economic ecology. Many scholars have carried out research on ecosystem services from different perspectives [25–30]. For example, Costanza et al. [7] established the equivalent factor table of global ecosystem service value based on equilibrium value theory and effect value theory [31,32]. Based on the research results by Costanza et al., Chinese scholars evaluated six services of the terrestrial ecosystem in China [33]. Chen et al. [34] carried out the value accounting on ES and benefits in China. Based on China's national conditions, Xie [35] revised the equivalent factor table through biomass parameters and developed the equivalent factor table of China's terrestrial ecosystem service value. This provided a calculation basis for scholars to study the value of regional ecosystem services in China [31,36]. The economic value of natural grassland in China, QTP, and Qinghai Province grassland was evaluated by Xie [37] and Lu [38]. At the same time, the impact of land use change on ESV was assessed in the karst area [39], China's Hengduan Mountains [40], the Three-river headwaters region Plain [41], Haibin [42], Qinghai-Tibet plateau [43], and Yangtze river [44] in China. Furthermore, more studies were performed in Mediterranean catchments [45], South Africa [46], Germany [47], the Andassa Watershed of Ethiopia [48], the central highlands of Ethiopia [49], and Bordeaux, France [50].

ES and LUCC are critical global challenges, influencing biodiversity, climate regulation, and sustainable development. Studies worldwide have highlighted the profound impacts of LUCC on ecological balance and ES provision. As an ecologically significant region, Qinghai Province on the Qinghai-Tibet Plateau plays a vital role in addressing these challenges, serving as a representative case for understanding the interactions between LUCC and ES in fragile environments. This study employs advanced methodologies, including land use dynamic degree and sensitivity coefficients, to analyze these dynamics, offering insights that extend beyond China. By bridging regional ecological challenges with global sustainability goals, the research provides valuable contributions to biodiversity conservation, sustainable land management, and the global discourse on ecosystem resilience.

Qinghai Province is a typical ecologically fragile region, which is representative of similar plateau environments with limited data. The characteristics of land use change are not obvious, and the dynamic changes in Ecosystem Service Value (ESV) remain unclear. To address these scientific issues, this study grapples with (1) land use dynamics during 2000 to 2018 in Qinghai Province; (2) clarifying ESV variations in response to LUCC changes; and (3) assessing the sensitivity of ESV to LUCC across various periods. By deeply studying the changes in ecosystem service value and their influencing factors in Qinghai Province, we can provide a scientific basis for formulating more effective ecological protection policies, as well as providing both theoretical and practical value for the sustainable use of resources in Qinghai Province.

2. Materials and Methods

2.1. Study Area

Qinghai Province is located in northwest China, at the junction of the Loess Plateau and QTP (between 31°39′–39°19′ N, 89°35′–103°04′ E). It is the fourth largest province in China, with an area of over 71.66104 × 10^4 km^2, spanning 1200 km from east to west and 800 km from north to south. The elevation in the province ranges from 1776 m to 6824 m (Figure 1). Qinghai Province has complex terrain and diversified landforms, with more than four-fifths of the area located on the plateau. Qinghai Province is characterized by a continental plateau climate with low temperatures (annual mean temperature from −5.6 °C to −8.6 °C), large diurnal temperature difference, less rainfall (mean annual precipitation from 176 mm to 764.4 mm), long duration of sunshine (about 2, 300~3, 600 h), and intense solar radiation (total radiation from 585.2 kJ/cm^2 to 739.86 kJ/cm^2).

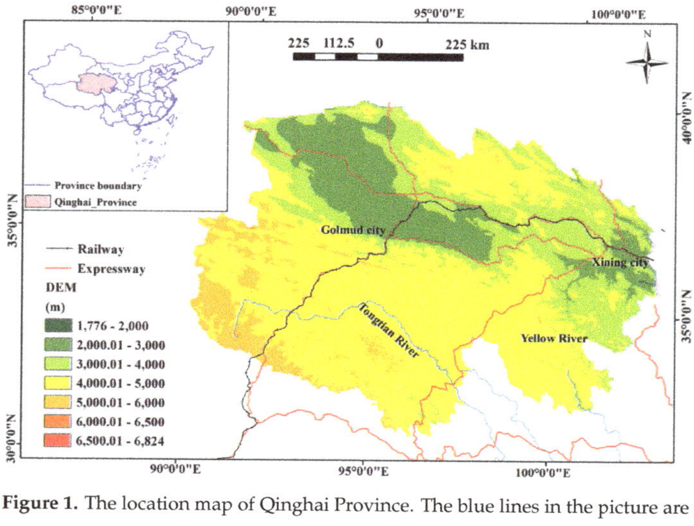

Figure 1. The location map of Qinghai Province. The blue lines in the picture are all rivers.

2.2. Data Sources

The land use/land cover dataset (vector) used in this study was developed by the Resources and Environment Data Centre of the Chinese Academy of Sciences (CAS) and provided at a scale of 1:100,000 in 2000, 2010, and 2018 [51,52]. The dataset is based on Landsat Thematic Mapper (TM), Enhanced Thematic Mapper (ETM+), and Operational Land Imager (OLI) remote sensing data for China (https://www.resdc.cn). The linear contrast stretching and histogram equalization were used to help identify ground control points in the rectification to a common Albers coordinate system based on 1:100,000 topographic maps of Qinghai Province. To guarantee the consistency and accuracy of data processing, the land use classification was conducted through visual interpretation and compared with Google Earth. After a random sample check, it was observed that the overall interpretation accuracy for the dataset is more than 84%.

2.3. Methods

2.3.1. LUCC Analysis Method

We adopted land use dynamic degree (K) (Equation (1)) and land use transfer matrix to analyze the characters of LUCC. The change range of land use types is usually expressed by a single land use dynamic degree in order to measure the change speed of a certain land use type in the study area [52]. The land transfer matrix not only quantitatively shows the transformation of different land use types but also reveals the transfer rate of different land use types. The land use transfer matrix is obtained in two consecutive steps. First, analysis tools-overlay-intersect in ArcGIS 10.6 to obtain the transfer data, and then the data are exported into PivotTable in Excel.

$$K = \frac{U_b - U_a}{U_a} \times \frac{1}{T} \times 100\% \tag{1}$$

where K is the dynamic degree of a certain land use type in the study period; U_a and U_b are the area of a kind of land use at the beginning and the end of the study period, respectively; T is the length of the study period.

Biodiversity is an important prerequisite for measuring ES. The combined effect of land use and climate change is an important driving force for biodiversity reduction. Richness and evenness are two components of diversity, referring to the number of patches present and the distribution of area among different types, respectively. Since biodiversity is a key factor in determining the capacity of ecosystems to provide services, Shannon diversity index (SHDI) serves as a useful proxy for assessing how land use changes and climate impacts may affect ESV. The richness component of SHDI, which measures the diversity of land cover types or patches, is particularly relevant in understanding how changes in landscape composition influence the provision of ecosystem services. A higher SHDI value suggests a more diverse and potentially more resilient ecosystem, which is linked to a greater capacity for providing a variety of ecosystem services. McGarigal and Marks [53] noted that the SHDI is more sensitive to richness than evenness. SHDI is obtained by FRAGSTATS 4.2 version (http://www.umass.edu/landeco/research/fragstats/fragstats.html, accessed on 1 May 2023). SHDI, an index of landscape diversity metrics, is a popular measure of diversity in community ecology, which is computed as (Equation (2)):

$$\text{SHDI} = -\sum_{i}^{n} P_i \ln P_i \tag{2}$$

where n is the number of landscape types, P_i is the proportion of landscape type i.

SHDI equals zero when the landscape contains only one landscape (no diversity). The value increases as the number of different landscapes increases, and the proportional distribution of area among landscapes becomes more equitable.

The equivalent factor method developed by Xie et al. [54] is adopted to calculate the economic values of ES. The method included a standard equivalent factor (Table 1) and an equivalent coefficients table (Table 2). The economic value of the natural grain output from 1 ha of farmland was used as the standard equivalent factor, and it was estimated at CNY 1191.08 per ha based on the average value from 2000 to 2018. The equivalent coefficients table is the valuation weight for each ecosystem service [43]. A certain type of ESV (Equation (3)) is computed as the product of the corresponding equivalent coefficients and the equivalent factor. Land use was divided into eight categories: 1—farmland, 2—woodland, 3—grassland, 4—water, 5—built-up land, 6—unused land, 7—wetland, and 8—glacier and snow. Built-up land was not considered by Xie et al. (2017) [54] and its ESV is assigned 0.

$$ESV_f = \sum_k \left(A_k \times VC_{kf} \right) \quad (3)$$

where ESV is the ecosystem service value (CNY); ESV_f is the value of ecosystem service f (CNY); A_k is ecosystem service value coefficient per unit area of land use k; VC_{kf} is the value type of ecosystem services corresponding to k (CNY. $ha^{-1} \cdot a^{-1}$).

Table 1. Response coefficient of land use to policies [54].

Project	Farmland	Woodland	Grassland	Wetland	Built-Up Land	Glacier and Snow	Water	Unused Land
Air quality regulation	3.5	0.8	0.5	1.8	0	0.18	0	0.4
Climate regulation	2.7	0.9	0.89	17.1	0	0.54	0.46	0.45
Soil fertility maintenance	3.2	0.8	0.6	15.5	0	7.13	20.38	0.415
Water conservation	3.9	1.95	1.46	1.71	0	0	0.01	0.985
Waste treatment	1.31	1.31	1.64	18.18	0	0.16	18.18	0.66
Biodiversity conservation	3.26	1.09	0.71	2.5	0	0.01	2.49	0.715
Food	0.1	0.3	1	0.3	0	0	0.1	0.155
Material	2.6	0.05	0.1	0.07	0	0	0.01	0.025
Cultural services	1.28	0.04	0.01	5.55	0	0.09	4.34	0.025

Table 2. Ecological system service value of land use types per unit area in Qinghai Province (unit: CNY·ha·a^{-1}).

Service Type	Farmland	Woodland	Grassland	Wetland	Built-Up Land	Glacier and Snow	Water	Unused Land
Air quality regulation	4168.78	952.864	595.54	2143.944	0	214.3944	0	476.432
Climate regulation	3215.916	1071.972	1060.061	20,367.47	0	643.1832	547.8968	535.986
Soil fertility maintenance	3811.456	952.864	714.648	18,461.74	0	8492.4	24,274.21	494.2982
Water conservation	4645.212	2322.606	1738.977	2036.747	0	0	11.9108	1173.214
Waste treatment	1560.315	1560.315	1953.371	21,653.83	0	190.5728	21,653.83	786.1128
Biodiversity conservation	3882.921	1298.277	845.6668	2977.7	0	11.9108	2965.789	851.6222
Food	119.108	357.324	1191.08	357.324	0	0	119.108	184.6174
Material	3096.808	59.554	119.108	83.3756	0	0	11.9108	29.777
Cultural services	1524.582	47.6432	11.9108	6610.494	0	107.1972	5169.287	29.777

2.3.2. CS Analysis Method

Sensitivity analysis aims to discuss the responses of sensitive subjects to sensitive factors. The factors pushing the sensitive subject to respond are defined as the sensitive factors. According to the definition, the CS (Equation (4)) is used to analyze the sensitivity of ESV to the VC over time [55]. CS refers to the degree of response of ESV to a 1% chance of VC. After revision, it is formulated as:

$$CS = \left| \frac{(ESV_j - ESV_i) / ESV_i}{(VC_{jk} - VC_{ik}) / VC_{ik}} \right| \quad (4)$$

where ESV_i and ESV_j are the initial and adjusted EVAs, respectively; VC_{ik} and VC_{jk} are the EVS coefficients (unit: CNY) before and after the adjustment, respectively; and k is the kth land use type.

It is obvious that if a small change of VC resulted in a significant ESV change, then the elasticity is large. So, when CS < 1, this indicates that the total ESV was relatively insensitive to variations in the equivalence value coefficients, the values of ecosystem services exhibited limited elasticity in comparison to the ecological value coefficients, and the results are credible. The greater the value of CS, the more the ESV is sensitive to the VC change and the more critical the accuracy of the ecosystem VC when assessing ESV.

3. Results

Significant changes in land use/land cover have taken place in Qinghai Province between 2000 and 2018 (Tables 3 and 4; Figures 2–4). The increase in the built-up land is the largest, while the woodland change is the least. In general, farmland, grassland, water, and wetland increased, while woodland and unused land decreased. The unused land decreased by 23,520 km^2. The water area showed a significant increase of 1729 km^2, and the dynamic degree was 12.52%. The woodland decreased by a total of 175 km^2 with a dynamic degree of −0.62%. The dynamic degree of grassland was −0.22% in 2000–2010, 5.62% in 2010–2018, and 5.39% in 2000–2018.

Table 3. Dynamic degree of land use type change in Qinghai Province.

Land Use Type	Farmland	Woodland	Grassland	Wetland	Built-Up Land	Glacier	Water	Unused Land
2000–2010	0.45%	−0.08%	−0.22%	−0.19%	36.60%	−0.56%	4.16%	−0.02%
2010–2018	3.44%	−0.54%	5.62%	2.82%	−3.33%	7.16%	8.03%	−9.13%
2000–2018	3.9%	−0.62%	5.39%	2.63%	32.05%	6.56%	12.52%	−9.15%

Table 4. Land use transition matrix in Qinghai Province from 2000 to 2010.

		2010 (km^2)								Total
		1	2	3	4	5	6	7	8	
2000 (km^2)	1	7862.71	8.12	97.39	14.25	36.58	4.59	1.45		8025.08
	2	14.16	25,862.23	424.77	1.66	0.03	30.83	4.66		26,338.34
	3	166.42	296.21	379,291.84	158.23	90.12	1415.39	35.81	0.20	381,454.22
	4	0.62	0.031	20.12	13,417.03	21.63	39.55	63.94		13,562.92
	5	3.31	0.39	0.04	0.62	893.46	1.63			903.04
	6	13.45	23.52	1193.60	411.76	111.83	255,503.99	319.51	3.82	257,581.48
	7	12.31	0.60	18.68	159.50	110.70	94.85	23,922.41		24,306.87
	8			0.19			25.92		4435.94	4462.05
Total		8060.79	26,191.10	381,050.23	14,163.05	1264.36	257,116.74	24,347.78	4439.95	716,634.00

Note: 1 farmland, 2 woodland, 3 grassland, 4 water, 5 built-up land, 6 unused land, 7 wetland, 8 glacier.

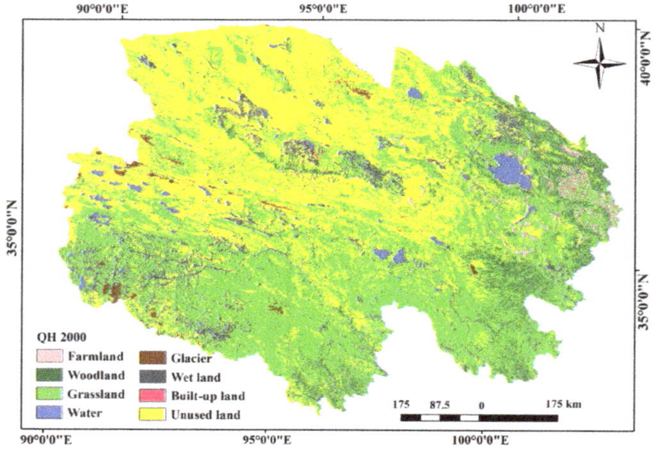

Figure 2. Land use/land cover in the Qinghai Province in 2000.

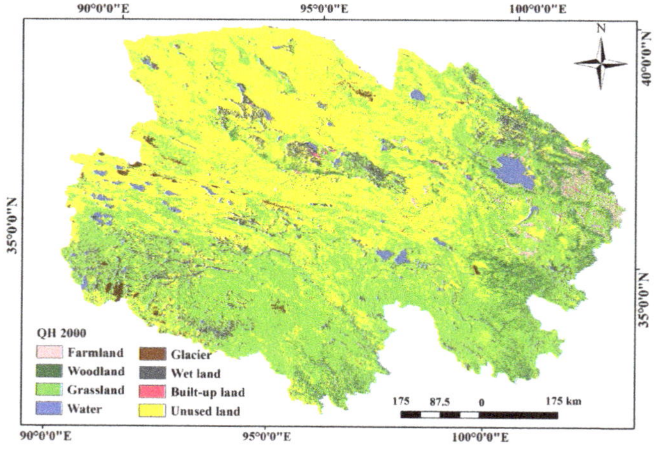

Figure 3. Land use/land cover in the Qinghai Province in 2010.

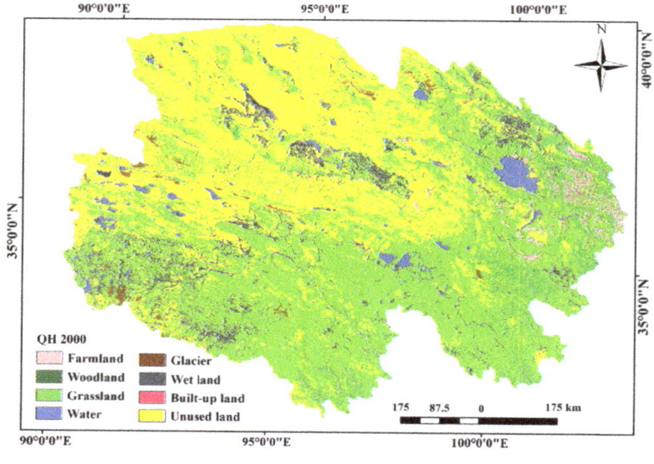

Figure 4. Land use/land cover in the Qinghai Province in 2018.

Tables 4–6 show the land use transition matrix from 2000 to 2018. It can be noticed that the land use/land cover of Qinghai Province has experienced significant change. From 2000 to 2010, 422.77 km^2 of woodland was transformed into grassland, and the grassland decreased by 820 km^2. Furthermore, 1415.39 km^2 of grasslands was transformed into unused land, and 1193.60 km^2 of unused land was transformed into grasslands (Table 4). In addition, 14.25 km^2 of farmland was submerged by water between 2000 and 2010.

Table 5. Land use transition matrix in Qinghai Province from 2010 to 2018.

		2018 (km^2)								Total
		1	2	3	4	5	6	7	8	
	1	4671.68	264.08	250,038	81.37	319.90	135.33	80.57		8053.31
	2	317.83	13,064.68	1,168,963.92	82.00	40.91	990.22	180.70	2.31	26,368.29
	3	2687.53	11,381.79	31,780,774	1449.24	352.84	38,410.61	8249.05	390.32	380,729.11
2010	4	73.99	79.26	124,326	11,707.57	19.33	645.56	770.41	0.88	14,540.26
(km^2)	5	275.60	40.00	29,156.84	125.84	280.67	1870.62	62.26		1263.55
	6	219.47	1218.93	58,95,534.13	1419.58	119.08	189,714.34	4149.81	1095.20	256,891.75
	7	65.10	190.91	818,891.81	673.01	51.30	3797.19	11,342.15	34.78	24,343.37
	8		3.69	32,764.61	0.39		845.31	28.45	3238.50	4444.00
Total		8311.20	26,243.38	40,100,452.32	15,539.04	1184.07	234,726.24	24,863.48	4762.08	716,634.00

Table 6. Land use transition matrix in Qinghai Province from 2000 to 2018.

		2018 (km^2)								
		1	2	3	4	5	6	7	8	Total
	1	5372.05	266.75	2524.00	88.17	338.34	130.91	80.40		8800.61
	2	323.96	12,934.85	11,735.60	82.83	38.80	990.87	181.09	2.31	26,290.31
	3	2781.23	11,294.21	317,786.54	1541.55	364.80	38,699.24	8274.64	391.09	381,133.31
2000	4	69.45	77.09	1090.08	11,036.38	20.06	571.97	691.82	0.88	13,557.75
(km^2)	5	25,776.02	3970.56	256.32	51.01	207.27	80.98	9.19		902.23
	6	22,482	121,381.84	590,7341	1677.00	150.10	189,639.98	4283.64	109,284.20	257,355.62
	7	64.94	188.25	8210.34	679.21	64.64	3746.01	11,314.17	34.78	24,302.35
	8		3.71	328.11	0.37		844.35	28.45	3086.82	4291.82
Total		9094.21	26,018.39	401,004.41	15,156.52	1184.02	234,704.32	24,863.41	4608.73	716,634.00

Farmland, grassland, wetland, and water increased by different degrees from 2010 to 2018. In total, 58,955.34 km^2 of unused land changed into grassland, while 38,410.61 km^2 of grassland changed to unused land. About 3797.19 km^2 of wetland turned into unused land, whereas 4149.81 km^2 of unused land turned into wetland. This suggests that the ecological environment is developing in a good direction in Qinghai Province.

From 2000 to 2018, except for the decrease in woodland and unused land, the other six land use types increased (Table 6). In total, 11,294.21 km^2 of grassland transformed into woodland, and 11,735.59 km^2 of woodland transformed into grassland. In 2000, the area of water was 13,557.75 km^2 and increased to 15,156.52 km^2 in 2018. In total, 1677.00 km^2 of unused land and 1541.55 km^2 of grassland transformed into water between 2000 and 2018.

The SHDI of landscape metrics of Qinghai Province decreased from 2000 to 2018 (Table 7). At first, it increased slightly from 2000 to 2010 but decreased significantly in 2010–2018. This result implies that the degree of fragmentation decreased, and that the complexity and stability of the ecosystem became worse. Accordingly, the land use types in Qinghai Province are developing in the direction of regularization and orderliness.

Table 7. Evolution of Landscape Diversity in Qinghai Province (SHDI).

Year	2000	2010	2018
SHDI	2.2350	2.24	2.1907

The ESV of Qinghai Province was CNY 750.77 billion in 2000 and increased to CNY 772.4 billion in 2018, with an increase of CNY 21.63 billion (Table 8). In 2018, supporting services, including water conservation, waste treatment, and biodiversity conservation, accounted for the largest percentage with over CNY 379.94 billion. Regulating services (including air quality regulation, climate regulation, and soil fertility maintenance) and provisioning services (including food and material) made the second and third largest contributions, accounting for over CNY 301.42 billion and over CNY 62.42 billion, respectively. The value of cultural service was about CNY 28.599 billion.

Table 8. The economic value of different types of ecosystem services.

Service Type	Ecosystem Services Value (Billion CNY)					
	2000	Proportion	2010	Proportion	2018	Proportion
Air quality regulation	46.61	6.2086%	46.56	6.1846%	46.98	6.0818%
Climate regulation	114.26	15.2194%	114.1	15.1563%	116.77	15.1178%
Soil fertility maintenance	131.56	17.5232%	132.79	17.6372%	137.68	17.8253%
Water conservation	111.58	14.8618%	111.43	14.8004%	112.61	14.5789%
Waste treatment	186.75	24.8747%	187.72	24.9337%	1941.51	25.1361%
Biodiversity conservation	72.64	9.6760%	72.74	9.6611%	73.18	9.4749%
Food	51.90	6.9127%	51.80	6.8808%	53.93	6.9821%
Material	8.22	1.0949%	8.22	1.0921%	8.50	1.1003%
Cultural services	27.24	3.6285%	27.51	3.6538%	28.60	3.7027%

The total service value of various ecosystems (farmland, grassland, water, glacier, snow, and wetland) showed an increase of varying degrees. The increase in service value of some land use types was the main factor leading to the total ESV increase. Table 9 demonstrates that the service value of the grassland ecosystem accounts for more than 41% of the total ESV. For example, the service value of the grassland ecosystem, with over CNY 327.22 billion, accounted for the largest percentage in 2018. The services value of wetland and unused land made the second and third largest contributions and accounted for over 26.1% and 13.7% of the total ESV, respectively. The service value of Glacier had the lowest ESV with about 0.5%.

Table 9. The ecosystem service value provided by different types of land use (billion CNY).

Land Use Types	Farmland	Woodland	Grassland	Wetland	Glacier	Water	Unused Land
2018	22.32	24.29	327.23	202.31	4.60	85.08	106.57
proportion	2.8896%	3.1441%	42.3651%	26.1927%	0.5955%	11.0153%	13.7976%
2010	21.58	24.42	309.81	196.76	4.29	78.76	117.28
proportion	2.8659%	3.2430%	41.1498%	26.1334%	0.5702%	10.4608%	15.5769%
2000	21.48	24.44	310.49	197.13	4.32	75.62	117.30
proportion	2.8612%	3.2548%	41.3560%	26.2570%	0.5750%	10.0717%	15.6242%

Table 10 shows that all CSs are less than 1. The maximum value of CS is for grassland with values between 0.4526 and 0.466. The minimum value of CS is for glaciers with values between 0.0063 and 0.0066. It suggests that the value of ecosystem services is inelastic to the coefficient of ecosystem. That is, the estimated value of ecosystem service is suitable, and the research results are credible for Qinghai Province.

Table 10. Coefficient of sensitivity of ecosystem service value in Qinghai Province.

Land Use Type	Farmland	Woodland	Grassland	Wetland	Built-Up Land	Glacier	Water	Unused Land
				(VC ± 1%)				
2018	0.0318	0.0346	0.4660	0.2881	0	0.0066	0.1212	0.1518
2010	0.0315	0.0357	0.4526	0.2875	0	0.0063	0.1151	0.1713
2000	0.0315	0.0358	0.4549	0.2888	0	0.0063	0.1108	0.1719

4. Discussion

4.1. Driving Forces of LUCC

Ecological environmental planning of Qinghai Province occupies a pivotal strategic position in China and has become a key area of national ecological protection and construction. Economic development strategies in Qinghai Province have brought new opportunities and challenges. Since the Reform and Opening up, the fast development of the agricultural economy and urbanization has improved the local economic level but has also destroyed the original ecological environment. LULC is the direct interaction between humans and natural ecosystems. Since 2007, China has given priority to building an "Ecological Civilization" by striving for green, circular, and low-carbon development. Qinghai Province, as one of the ecologically fragile regions, has taken a series of measures to address environmental and development problems [56].

The pattern of LUCC can be attributed to different driving factors. With the tremendous progress in social and economic development as well as population growth, the size and scale of urbanization and industrialization are unprecedented [57], which resulted in the decrease of woodland and unused land [58]. The increase in precipitation caused by climate warming and human activities influenced the formation of new lakes and the expansion of water bodies [43,58–63]. Due to the combined effect of human activity and climate change, the area of grassland increased in Qinghai Province [59]. Farmland area increased steadily due to the implementation of farmland protection policies such as "Dynamic Balance of Total farmland", "Occupancy and Balance System of farmland", and "Land use Control".

4.2. Implications for Ecosystem Management

Although SHDI showed a slight decline during 2000–2018 (Table 7), the values were all greater than 1, which indicates that the spatial distribution of landscape patterns is more balanced due to landscape planning and design. This is also one of the achievements of ecosystem management. CS is a simple tool used to measure how changes in ESV respond to specific factors. It helps identify unusual changes in different ecosystems and guides actions needed for better ecosystem management [64]. Results showed that the CS change of unused land and water was larger than for other ecosystems. The CS of grassland is the greatest (Table 10), followed by wetland and unused land. It suggests that if the area of grassland, wetland, and unused land changes substantially, it would seriously affect the total ecosystem services value and the regulating, provisioning, supporting, and cultural services [4,65–67]). Accordingly, natural reserves should be established through an adaptive strategy to protect the grassland, wetland, and unused land.

In the context of biodiversity in Qinghai Province, Table 2 highlights that while production forests and meadows have low biodiversity due to strict species management and focus on economic output, agricultural production areas, despite being dominated by monocultures and the extensive use of pesticides, exhibit a higher species richness. This higher species count, however, does not necessarily reflect native or natural species characteristic of the region. Rather, these areas support a mix of native and non-native species, including weeds and other organisms that thrive in disturbed environments. The biodiversity in agricultural areas is, therefore, not indicative of the region's natural biodiversity but rather the result of human influence on the landscape. To provide further clarity, the paper will discuss changes in biodiversity over time and the extent to which native species are being replaced or maintained in these areas.

4.3. The Factors Affecting the Distribution of ESV

In our ESV framework, some of the values, particularly for farmland's role in air regulation and water conservation, may seem counterintuitive. Farmland plays a modest role in air regulation through carbon sequestration and microclimate moderation, which justifies its value for air regulation, though it is not as significant in natural ecosystems like forests. Similarly, while farmland does not directly contribute to water purification, it can aid in water retention in the soil and prevent surface runoff, albeit to a lesser extent than wetlands or forests. For water conservation, the value of farmland reflects the fact that agricultural areas, especially those with irrigation, often consume more water than they conserve, which is why the value may appear low. Regarding glaciers, the high value for soil fertility maintenance is an indirect result of glacial erosion processes that contribute to soil formation over long periods. While glaciers do not actively contribute to soil fertility in the short term, their impact on the geological processes that shape fertile soils is reflected in the ecosystem service value.

Through our study, it can be found that the ESV in the study area shows an increasing trend overall. The distribution of ESV is influenced by a variety of factors, both natural and anthropogenic. Understanding these factors is crucial for interpreting the spatial variations in ESV and for formulating effective ecosystem management strategies.

The distribution of ESV is shaped by various natural and human factors. Different land use types play a crucial role in determining ESV, with grasslands and wetlands typically associated with highly regulating and supporting services, such as carbon sequestration and water purification, leading to higher ESV values. In contrast, urbanized areas generally exhibit lower ESV due to the conversion of natural landscapes into built environments, which results in the loss of biodiversity and ecosystem functions. Climate also significantly influences the availability and distribution of ecosystem services. Regions with favorable climatic conditions, such as higher precipitation and moderate temperatures, tend to support more productive ecosystems, thereby providing higher ESV. In contrast, arid or extreme temperature zones often exhibit lower ESV due to limited vegetation and ecosystem services. Elevation further impacts ESV, as higher elevation areas, like those in the Qinghai-Tibet Plateau, support unique ecosystems that provide essential services such as water regulation and biodiversity conservation. These areas typically show higher ESV due to their ecological importance, despite limited human intervention. Additionally, soil fertility and vegetation cover are critical factors affecting ESV distribution. Regions with fertile soils and abundant vegetation are more likely to support a high ESV, as these ecosystems provide a range of vital services, from soil formation to habitat provision. Conversely, areas with degraded soils or sparse vegetation often show lower ESV due to diminished ecological productivity. Together, these factors underscore the complexity of ESV distribution and highlight the need for targeted conservation strategies tailored to regional environmental characteristics.

4.4. The Accuracy of Results

The accuracy of the obtained results is determined by the quality of the LULC data and the unit ESV of each LULC type [68]. When using Landsat data for land use/land cover classification, there would be omission and commission [69,70]. Therefore, without explicitly considering the uncertainty of classification, it is possible to obtain a rather inflated estimate of LUCC. Additionally, the compatibility, comparability, and interoperability of heterogeneous land cover datasets are greatly limited due to the lack of land cover classification standards [71]. Four available land use/land cover datasets at spatial resolutions from 30 m to 250 m are used by scholars in China. They include two subsets of the global land cover maps at 30 m × 30 m spatial resolution [72,73], the China Cover dataset at 250 m × 250 m, and the national land use/land cover dataset used in this study. Compared with the other three data sets, the national land use/land cover dataset has a longer time series and higher accuracy, with the accuracy rate of field verification being more than 94% [74,75].

The accuracy of results depends on land use data quality, with the national dataset offering higher accuracy and a longer time series. Xie et al.'s ecosystem service values are more suited to Qinghai's context compared to Costanza et al.'s estimates.

5. Conclusions

The combination of LUCC and unit value transfer is the most practical and effective method to assess ESV in provincial and even larger regions where fieldwork is difficult and costly due to their harsh natural environment and lack of data. Qinghai Province is a well-valued region for providing multiple vital ecosystem services. Accordingly, we assessed its ESV changes due to LUCC in the period 2000–2018. The results are summarized as follows:

(1) The total ecosystem service value (ESV), encompassing regulating, provisioning, supporting, and cultural services, increased from CNY 750.77 billion in 2000 to CNY 772.4 billion in 2018. This growth is attributed to a significant expansion in water and farmland areas, coupled with reductions in unused land and glacier regions, driven by the impacts of climate change and the implementation of eco-environmental protection policies.

(2) The analysis of CS to ecosystem service value coefficient in three periods showed that the response of ESV changes is inelastic to LUCC, yet CS can still be used to indicate changes in different land use and assist in the specific ecosystem management.

(3) The accuracy of these results is determined by the quality of the LUCC and the unit ecosystem service value of each land use/land cover type. The improvement of LUCC and increasing the number of empirical studies and contrastive analysis are necessary in order to obtain a more accurate unit ecosystem service value in the future.

Author Contributions: Conceptualization, L.L. and W.N.; methodology, L.L. and X.L.; data curation, X.L.; software, L.L. and M.E.-S.A.; validation, L.L., A.E. and A.A.; investigation, L.L. and Z.H.; writing—original draft preparation, L.L., Y.Z. and A.A.; writing—review and editing, W.N., Y.Z. and K.Z. All authors have read and agreed to the published version of the manuscript.

Funding: This research was funded by the National Foreign Expert Individual Human Project (Category H) (H20240400); National Natural Science Foundation of China (Grant No. 42261144749, 42377158); International Science and Technology Cooperation Program of Shaanxi Province (Grant No. 2024GH-ZDXM-24), Shaanxi Province Agricultural science and technology 114 public welfare platform to serve rural revitalization practical technical training (Grant No. 2024NC-XCZX-06); Second Tibetan Plateau Scientific Expedition and Research Program (STEP), grant number 2019QZKK0905; National Scientific Research Project Cultivation Fund of Huanghuai University (NO. XKPY-2023023), and the Zhumadian Key R&D Special Project (ZMDSZDZX2023006).

Institutional Review Board Statement: Not applicable.

Informed Consent Statement: Not applicable.

Data Availability Statement: The original contributions presented in this study are included in the article material, and further inquires can be directed to the corresponding author.

Acknowledgments: We sincerely thank the editor and all reviewers for their valuable comments and feedback. We also thank the National Tibetan Plateau Data Center for providing the additional dataset.

Conflicts of Interest: The authors declare no conflicts of interest.

References

1. Huo, A.; Zhao, Z.; Luo, P.; Zheng, C.; Peng, J.; Abuarab, M.E.-S. Assessment of spatial heterogeneity of soil moisture in the critical zone of gully consolidation and highland protection. *Water* **2022**, *14*, 3674. [CrossRef]
2. Fan, Y.; Fang, C. Evolution process and obstacle factors of ecological security in western China, a case study of Qinghai province. *Ecol. Indic.* **2020**, *117*, 106659. [CrossRef]
3. Huo, A.; Huang, Z.; Cheng, Y.; Van Liew, M.W. Comparison of two different approaches for sensitivity analysis in Heihe River basin (China). *Water Supply* **2020**, *20*, 319–327. [CrossRef]
4. Li, X.; Gao, J.; Zhang, J.; Sun, H. Natural and anthropogenic influences on the spatiotemporal change of degraded meadows in southern Qinghai Province, West China: 1976–2015. *Appl. Geogr.* **2018**, *97*, 176–183. [CrossRef]
5. Huo, A.-D.; Dang, J.; Song, J.-X.; Chen, X.H.; Mao, H.-R. Simulation modeling for water governance in basins based on surface water and groundwater. *Agric. Water Manag.* **2016**, *174*, 22–29. [CrossRef]
6. Kang, B.; Shao, Q.; Xu, H.; Jiang, F.; Wei, X.; Shao, X. Research on grassland ecosystem service value in China under climate change based on meta-analysis: A case study of Qinghai province. *Int. J. Clim. Chang. Strateg. Manag.* **2020**, *12*, 617–637. [CrossRef]
7. Costanza, R.; d'Arge, R.; De Groot, R.; Farber, S.; Grasso, M.; Hannon, B.; Limburg, K.; Naeem, S.; O'neill, R.V.; Paruelo, J. The value of the world's ecosystem services and natural capital. *Nature* **1997**, *387*, 253–260. [CrossRef]
8. Sánchez, M.L.; Cabrera, A.T.; del Pulgar, M.L.G. The potential role of cultural ecosystem services in heritage research through a set of indicators. *Ecol. Indic.* **2020**, *117*, 106670. [CrossRef]
9. Qin, C.; Tang, Z.; Chen, J.; Chen, X. The impact of soil and water resource conservation on agricultural production-an analysis of the agricultural production performance in Zhejiang, China. *Agric. Water Manag.* **2020**, *240*, 106268. [CrossRef]
10. Millennium Ecosystem Assessment. *Ecosystems and Human Well-Being: A Framework for Assessment*; Millennium Ecosystem Assessment: Washington, DC, USA, 2003.
11. Sala, O.E.; Stuart Chapin, F.; Armesto, J.J.; Berlow, E.; Bloomfield, J.; Dirzo, R.; Huber-Sanwald, E.; Huenneke, L.F.; Jackson, R.B.; Kinzig, A. Global biodiversity scenarios for the year 2100. *Science* **2000**, *287*, 1770–1774. [CrossRef]
12. Preston, B.L.; Yuen, E.J.; Westaway, R.M. Putting vulnerability to climate change on the map: A review of approaches, benefits, and risks. *Sustain. Sci.* **2011**, *6*, 177–202. [CrossRef]
13. Shao, J.; Ni, J.; Wei, C.; Xie, D. Land use change and its corresponding ecological responses: A review. *J. Geogr. Sci.* **2005**, *15*, 305–328. [CrossRef]
14. Foley, J.A.; DeFries, R.; Asner, G.P.; Barford, C.; Bonan, G.; Carpenter, S.R.; Chapin, F.S.; Coe, M.T.; Daily, G.C.; Gibbs, H.K. Global consequences of land use. *Science* **2005**, *309*, 570–574. [CrossRef]
15. Song, X.-P.; Hansen, M.C.; Stehman, S.V.; Potapov, P.V.; Tyukavina, A.; Vermote, E.F.; Townshend, J.R. Global land change from 1982 to 2016. *Nature* **2018**, *560*, 639–643. [CrossRef]
16. Tian, Y.; Zhou, D.; Jiang, G. Conflict or Coordination? Multiscale assessment of the spatio-temporal coupling relationship between urbanization and ecosystem services: The case of the Jingjinji Region, China. *Ecol. Indic.* **2020**, *117*, 106543. [CrossRef]
17. Hu, M.; Wang, Y.; Xia, B.; Jiao, M.; Huang, G. How to balance ecosystem services and economic benefits?—A case study in the Pearl River Delta, China. *J. Environ. Manag.* **2020**, *271*, 110917. [CrossRef]
18. Tolvanen, A.; Kangas, K.; Tarvainen, O.; Huhta, E.; Jäkäläniemi, A.; Kyttä, M.; Nikula, A.; Nivala, V.; Tuulentie, S.; Tyrväinen, L. The relationship between people's activities and values with the protection level and biodiversity. *Tour. Manag.* **2020**, *81*, 104141. [CrossRef]
19. Wu, C.; Chen, B.; Huang, X.; Wei, Y.D. Effect of land-use change and optimization on the ecosystem service values of Jiangsu province, China. *Ecol. Indic.* **2020**, *117*, 106507. [CrossRef]
20. Liu, Y.; Hou, X.; Li, X.; Song, B.; Wang, C. Assessing and predicting changes in ecosystem service values based on land use/cover change in the Bohai Rim coastal zone. *Ecol. Indic.* **2020**, *111*, 106004. [CrossRef]
21. Ren, Y.; Lu, L.; Zhang, H.; Chen, H.; Zhu, D. Residents' willingness to pay for ecosystem services and its influencing factors: A study of the Xin'an River basin. *J. Clean. Prod.* **2020**, *268*, 122301. [CrossRef]
22. Constant, N.L.; Taylor, P.J. Restoring the forest revives our culture: Ecosystem services and values for ecological restoration across the rural-urban nexus in South Africa. *For. Policy Econ.* **2020**, *118*, 102222. [CrossRef]
23. Liang, X.; Jin, X.; Ren, J.; Gu, Z.; Zhou, Y. A research framework of land use transition in Suzhou City coupled with land use structure and landscape multifunctionality. *Sci. Total Environ.* **2020**, *737*, 139932. [CrossRef]

24. Shi, L.; Halik, Ü.; Mamat, Z.; Wei, Z. Spatio-temporal variation of ecosystem services value in the Northern Tianshan Mountain Economic zone from 1980 to 2030. *PeerJ* **2020**, *8*, e9582. [CrossRef]
25. De Groot, R.; Brander, L.; Van Der Ploeg, S.; Costanza, R.; Bernard, F.; Braat, L.; Christie, M.; Crossman, N.; Ghermandi, A.; Hein, L. Global estimates of the value of ecosystems and their services in monetary units. *Ecosyst. Serv.* **2012**, *1*, 50–61. [CrossRef]
26. Costanza, R.; De Groot, R.; Sutton, P.; Van der Ploeg, S.; Anderson, S.J.; Kubiszewski, I.; Farber, S.; Turner, R.K. Changes in the global value of ecosystem services. *Glob. Environ. Chang.* **2014**, *26*, 152–158. [CrossRef]
27. Viglizzo, E.F.; Paruelo, J.M.; Laterra, P.; Jobbágy, E.G. Ecosystem service evaluation to support land-use policy. *Agric. Ecosyst. Environ.* **2012**, *154*, 78–84. [CrossRef]
28. Fu, B.; Zhang, L.; Xu, Z.; Zhao, Y.; Wei, Y.; Skinner, D. Ecosystem services in changing land use. *J. Soils Sediments* **2015**, *15*, 833–843. [CrossRef]
29. Frélichová, J.; Vačkář, D.; Pártl, A.; Loučková, B.; Harmáčková, Z.V.; Lorencová, E. Integrated assessment of ecosystem services in the Czech Republic. *Ecosyst. Serv.* **2014**, *8*, 110–117. [CrossRef]
30. Bateman, I.J.; Harwood, A.R.; Mace, G.M.; Watson, R.T.; Abson, D.J.; Andrews, B.; Binner, A.; Crowe, A.; Day, B.H.; Dugdale, S. Bringing ecosystem services into economic decision-making: Land use in the United Kingdom. *Science* **2013**, *341*, 45–50. [CrossRef]
31. Ran, C.; Wang, S.; Bai, X.; Tan, Q.; Zhao, C.; Luo, X.; Chen, H.; Xi, H. Trade-offs and synergies of ecosystem services in southwestern China. *Environ. Eng. Sci.* **2020**, *37*, 669–678. [CrossRef]
32. Huo, A.; Peng, J.; Cheng, Y.; Zheng, X.; Wen, Y. Temporal characteristics of the rainfall induced landslides in the Chinese loess plateau (China). In *Recent Advances in Geo-Environmental Engineering, Geomechanics and Geotechnics, and Geohazards: Proceedings of the 1st Springer Conference of the Arabian Journal of Geosciences (CAJG-1), Tunisia 2018*; Springer: Berlin/Heidelberg, Germany, 2019; pp. 425–427.
33. Ouyang, Z.; Wang, X.; Miao, H. A primary study on Chinese terrestrial ecosystem services and their ecological-economic values. *Acta Ecol. Sin.* **1999**, *19*, 607–613.
34. Chen, Z.; Zhang, X. The value of ecosystem benefits in China. *Chin. Sci. Bull.* **2000**, *45*, 17–22.
35. Xie, G.; Zhang, Y.; Lu, C.; Zheng, D.; Cheng, S. Study on valuation of rangeland ecosystem services of China. *J. Nat. Resour.* **2001**, *16*, 47–53.
36. Song, W.; Deng, X. Land-use/land-cover change and ecosystem service provision in China. *Sci. Total Environ.* **2017**, *576*, 705–719. [CrossRef]
37. Xie, G.D. The economic evaluation of grassland ecosystem services in Qinghai-Tibet Plateau. *J. Mt. Sci.* **2003**, *21*, 50–55.
38. Lu, C.; Xie, G.; Xiao, Y.; Yu, Y. Ecosystem diversity and economic valuation of Qinghai-Tibet Plateau. *Acta Ecol. Sin.* **2004**, *24*, 2749–2755.
39. Hu, Z.; Wang, S.; Bai, X.; Luo, G.; Li, Q.; Wu, L.; Yang, Y.; Tian, S.; Li, C.; Deng, Y. Changes in ecosystem service values in karst areas of China. *Agric. Ecosyst. Environ.* **2020**, *301*, 107026. [CrossRef]
40. Wang, Y.; Dai, E.; Yin, L.; Ma, L. Land use/land cover change and the effects on ecosystem services in the Hengduan Mountain region, China. *Ecosyst. Serv.* **2018**, *34*, 55–67. [CrossRef]
41. Zheng, D.; Wang, Y.; Hao, S.; Xu, W.; Lv, L.; Yu, S. Spatial-temporal variation and tradeoffs/synergies analysis on multiple ecosystem services: A case study in the Three-River Headwaters region of China. *Ecol. Indic.* **2020**, *116*, 106494. [CrossRef]
42. Riao, D.; Zhu, X.; Tong, Z.; Zhang, J.; Wang, A. Study on land use/cover change and ecosystem services in Harbin, China. *Sustainability* **2020**, *12*, 6076. [CrossRef]
43. Jiang, W.; Lü, Y.; Liu, Y.; Gao, W. Ecosystem service value of the Qinghai-Tibet Plateau significantly increased during 25 years. *Ecosyst. Serv.* **2020**, *44*, 101146. [CrossRef]
44. Dai, X.; Wang, L.; Huang, C.; Fang, L.; Wang, S.; Wang, L. Spatio-temporal variations of ecosystem services in the urban agglomerations in the middle reaches of the Yangtze River, China. *Ecol. Indic.* **2020**, *115*, 106394. [CrossRef]
45. Boix-Fayos, C.; Boerboom, L.G.; Janssen, R.; Martínez-Mena, M.; Almagro, M.; Pérez-Cutillas, P.; Eekhout, J.P.; Castillo, V.; De Vente, J. Mountain ecosystem services affected by land use changes and hydrological control works in Mediterranean catchments. *Ecosyst. Serv.* **2020**, *44*, 101136. [CrossRef]
46. Anderson, S.J.; Ankor, B.L.; Sutton, P.C. Ecosystem service valuations of South Africa using a variety of land cover data sources and resolutions. *Ecosyst. Serv.* **2017**, *27*, 173–178. [CrossRef]
47. Jiang, W. Mapping ecosystem service value in Germany. *Int. J. Sustain. Dev. World Ecol.* **2018**, *25*, 518–534. [CrossRef]
48. Gashaw, T.; Tulu, T.; Argaw, M.; Worqlul, A.W.; Tolessa, T.; Kindu, M. Estimating the impacts of land use/land cover changes on Ecosystem Service Values: The case of the Andassa watershed in the Upper Blue Nile basin of Ethiopia. *Ecosyst. Serv.* **2018**, *31*, 219–228. [CrossRef]

49. Tolessa, T.; Senbeta, F.; Kidane, M. The impact of land use/land cover change on ecosystem services in the central highlands of Ethiopia. *Ecosyst. Serv.* **2017**, *23*, 47–54. [CrossRef]
50. Cabral, P.; Feger, C.; Levrel, H.; Chambolle, M.; Basque, D. Assessing the impact of land-cover changes on ecosystem services: A first step toward integrative planning in Bordeaux, France. *Ecosyst. Serv.* **2016**, *22*, 318–327. [CrossRef]
51. Liu, J.; Liu, M.; Zhuang, D.; Zhang, Z.; Deng, X. Study on spatial pattern of land-use change in China during 1995–2000. *Sci. China Ser. D Earth Sci.* **2003**, *46*, 373–384. [CrossRef]
52. Liu, J.; Zhang, Z.; Xu, X.; Kuang, W.; Zhou, W.; Zhang, S.; Li, R.; Yan, C.; Yu, D.; Wu, S. Spatial patterns and driving forces of land use change in China during the early 21st century. *J. Geogr. Sci.* **2010**, *20*, 483–494. [CrossRef]
53. McGarigal, K. *FRAGSTATS: Spatial Pattern Analysis Program for Quantifying Landscape Structure*; US Department of Agriculture, Forest Service, Pacific Northwest Research Station: Portland, OR, USA, 1995; Volume 351.
54. Xie, G.; Zhang, C.; Zhen, L.; Zhang, L. Dynamic changes in the value of China's ecosystem services. *Ecosyst. Serv.* **2017**, *26*, 146–154. [CrossRef]
55. Yue, Y.; Xue, L. Study on the dynamics of land use and ecosystem services value in Shaanxi Province. *J. China Agric. Univ.* **2020**, *25*, 20–30.
56. Ye, S.; Ge, Y.; Xu, S.; Ma, X. Measurement and Prediction of Coupling Coordination Level of Economic Development, Social Stability and Ecological Environment in Qinghai—Thoughts on Sustainable Societal Safety. *Sustainability* **2022**, *14*, 10515. [CrossRef]
57. Gu, C.; Hu, L.; Cook, I.G. China's urbanization in 1949–2015: Processes and driving forces. *Chin. Geogr. Sci.* **2017**, *27*, 847–859. [CrossRef]
58. Cui, X.; Graf, H.-F.; Langmann, B.; Chen, W.; Huang, R. Hydrological impacts of deforestation on the southeast Tibetan plateau. *Earth Interact.* **2007**, *11*, 1–18. [CrossRef]
59. Xu, W.; Ma, L.; Ma, M.; Zhang, H.; Yuan, W. Spatial–temporal variability of snow cover and depth in the Qinghai–Tibetan Plateau. *J. Clim.* **2017**, *30*, 1521–1533. [CrossRef]
60. Yao, T.; Thompson, L.; Yang, W.; Yu, W.; Gao, Y.; Guo, X.; Yang, X.; Duan, K.; Zhao, H.; Xu, B. Different glacier status with atmospheric circulations in Tibetan Plateau and surroundings. *Nat. Clim. Chang.* **2012**, *2*, 663–667. [CrossRef]
61. Sun, J.; Zhou, T.; Liu, M.; Chen, Y.; Shang, H.; Zhu, L.; Shedayi, A.A.; Yu, H.; Cheng, G.; Liu, G. Linkages of the dynamics of glaciers and lakes with the climate elements over the Tibetan Plateau. *Earth-Sci. Rev.* **2018**, *185*, 308–324. [CrossRef]
62. Dong, H.; Song, Y.; Zhang, M. Hydrological trend of Qinghai Lake over the last 60 years: Driven by climate variations or human activities? *J. Water Clim. Chang.* **2019**, *10*, 524–534. [CrossRef]
63. Chang, B.; He, K.-N.; Li, R.-J.; Sheng, Z.-P.; Wang, H. Linkage of climatic factors and human activities with water level fluctuations in Qinghai Lake in the northeastern Tibetan Plateau, China. *Water* **2017**, *9*, 552. [CrossRef]
64. Li, X.-Y.; Xu, H.-Y.; Sun, Y.-L.; Zhang, D.-S.; Yang, Z.-P. Lake-level change and water balance analysis at Lake Qinghai, west China during recent decades. *Water Resour. Manag.* **2007**, *21*, 1505–1516. [CrossRef]
65. Faccioli, M.; Czajkowski, M.; Glenk, K.; Martin-Ortega, J. Environmental attitudes and place identity as determinants of preferences for ecosystem services. *Ecol. Econ.* **2020**, *174*, 106600. [CrossRef]
66. Ren, Y.; Lü, Y.; Fu, B. Quantifying the impacts of grassland restoration on biodiversity and ecosystem services in China: A meta-analysis. *Ecol. Eng.* **2016**, *95*, 542–550. [CrossRef]
67. Wang, Y.; Sun, Y.; Wang, Z.; Chang, S.; Hou, F. Grazing management options for restoration of alpine grasslands on the Qinghai-Tibet Plateau. *Ecosphere* **2018**, *9*, e02515. [CrossRef]
68. Xu, X.; Jiang, B.; Tan, Y.; Costanza, R.; Yang, G. Lake-wetland ecosystem services modeling and valuation: Progress, gaps and future directions. *Ecosyst. Serv.* **2018**, *33*, 19–28. [CrossRef]
69. Sun, M.; Ma, W.; Yao, X.; Zhao, L.; Li, Z.; Qin, D. Evaluation and spatiotemporal characteristics of glacier service value in the Qilian Mountains. *J. Geogr. Sci.* **2020**, *30*, 1233–1248. [CrossRef]
70. Foody, G.M. Valuing map validation: The need for rigorous land cover map accuracy assessment in economic valuations of ecosystem services. *Ecol. Econ.* **2015**, *111*, 23–28. [CrossRef]
71. Song, X.-P. Global estimates of ecosystem service value and change: Taking into account uncertainties in satellite-based land cover data. *Ecol. Econ.* **2018**, *143*, 227–235. [CrossRef]
72. Yang, H.; Li, S.; Chen, J.; Zhang, X.; Xu, S. The standardization and harmonization of land cover classification systems towards harmonized datasets: A review. *ISPRS Int. J. Geo-Inf.* **2017**, *6*, 154. [CrossRef]
73. Gong, P.; Wang, J.; Yu, L.; Zhao, Y.; Zhao, Y.; Liang, L.; Niu, Z.; Huang, X.; Fu, H.; Liu, S. Finer resolution observation and monitoring of global land cover: First mapping results with Landsat TM and ETM+ data. *Int. J. Remote Sens.* **2013**, *34*, 2607–2654. [CrossRef]

74. Chen, J.; Chen, J.; Liao, A.; Cao, X.; Chen, L.; Chen, X.; He, C.; Han, G.; Peng, S.; Lu, M. Global land cover mapping at 30 m resolution: A POK-based operational approach. *ISPRS J. Photogramm. Remote Sens.* **2015**, *103*, 7–27. [CrossRef]
75. Ning, J.; Liu, J.; Kuang, W.; Xu, X.; Zhang, S.; Yan, C.; Li, R.; Wu, S.; Hu, Y.; Du, G. Spatiotemporal patterns and characteristics of land-use change in China during 2010–2015. *J. Geogr. Sci.* **2018**, *28*, 547–562. [CrossRef]

sustainability

MDPI

Article

Evaluation Index System for Thermokarst Lake Susceptibility: An Effective Tool for Disaster Warning on the Qinghai–Tibet Plateau, China

Lan Li [1,*], Yilu Zhao [2,*], Xuan Li [1], Wankui Ni [2] and Fujun Niu [3]

1 College of Civil Engineering and Architecture, Huanghuai University, Zhumadian 463000, China; lixuan9026@163.com
2 School of Water and Environment, Chang'an University, Xi'an 710054, China; niwankui@chd.edu.cn
3 College of Geological Engineering and Geomatics, Chang'an University, Xi'an 710054, China; niufujun@lzb.ac.cn
* Correspondence: lilan@huanghuai.edu.cn (L.L.); zhaoyilu1999@163.com (Y.Z.)

Abstract: In the context of global warming, landscapes with ice-rich permafrost, such as the Qinghai–Tibet Plateau (QTP), are highly vulnerable. The expansion of thermokarst lakes erodes the surrounding land, leading to collapses of various scales and posing a threat to nearby infrastructure and the environment. Assessing the susceptibility of thermokarst lakes in remote, data-scarce areas remains a challenging task. In this study, Landsat imagery and human–computer interaction were employed to improve the accuracy of thermokarst lake classification. The study also identified the key factors influencing the occurrence of thermokarst lakes, including the lake density, soil moisture (SM), slope, vegetation, snow cover, ground temperature, precipitation, and permafrost stability (PS). The results indicate that the most susceptible areas cover 19.02% of the QTP's permafrost region, primarily located in southwestern Qinghai, northeastern Tibet, and the Hoh Xil region. This study provides a framework for mapping the spatial distribution of thermokarst lakes and contributes to understanding the impact of climate change on the QTP.

Keywords: thermokarst lakes; susceptibility assessment; permafrost region; Qinghai–Tibet Plateau

1. Introduction

With the intensification of climate change and frequent human engineering activities, permafrost will inevitably lead to regional changes in the frozen soil environment; among the most significant manifestation is the extensive development of thermokarst lakes [1–3]. Thermokarst lakes are collapsed depressions of talik that result from the partial melting of ground ice or ice-rich permafrost, owing to natural and human factors; as a widely distributed melting disaster, it can make the ground temperature around engineering rise and lead to many adverse phenomena, such as deformation, retrogressive thaw slump, settlement, slope collapse and so on [4–10].

Field investigations and the analysis of historical remote sensing data on thermokarst lakes show a marked upward trend in the number and area of permafrost regions on the QTP [11]. From 1969 to 2019, there was a >0.1 ha increase in the number of thermokarst lakes (+158%) and their surface area (+123%) on the central Tibetan Plateau [12]. In North-Eastern Siberia, researchers have discovered that geomorphological and climatic factors promoted the increase in thermokarst lakes during 1990–2018 [13]. The increase in thermokarst lakes

https://doi.org/10.3390/su17041464

could lead to the release of carbon produced by the anaerobic decomposition of fossils and fresh organic matter previously stored in permafrost into the air [14–18]. Additionally, thermokarst lakes changes have a dramatic effect on water balance and the hydrologic cycle [19,20].

The formation of thermokarst lakes on the Qinghai–Tibet Plateau is not influenced by a single factor; rather, these factors interact with each other [21,22]. Higher ground temperatures may amplify the effects of precipitation and accelerate the melting of permafrost [23]. Similarly, areas with minimal vegetation are more susceptible to temperature changes, leading to faster permafrost degradation and lake expansion [24]. Additionally, human activities such as infrastructure development exacerbate these processes, accelerating permafrost thawing and thermokarst lake expansion.

Remote sensing technology can cover a wide geographical area and quickly obtain surface information, which is crucial for identifying potential thermokarst lake areas. Through remote sensing data from different bands, various parameters such as ground temperature, soil moisture, and vegetation coverage can be analyzed, and are closely related to the formation and expansion of thermokarst lakes. On-site investigation can verify and calibrate remote sensing data to ensure its accuracy and reliability. By combining remote sensing data with on-site survey data, a more comprehensive dataset can be constructed for training and optimizing evaluation models.

Susceptibility assessment lays a key foundation for subsequent research on disaster and risk assessment, and is of great significance for environmental protection, land use planning, engineering site selection, disaster prevention and reduction, etc. [2,25–27]. Susceptibility refers to the likelihood or probability of a particular event or condition occurring [2]. The determination of an evaluation index system, weights and evaluation method is crucial for obtaining accurate results. Liu et al. used Sentinel-2 images to analyze thermokarst lake boundary transitions, with the results demonstrating that the development of thermokarst lakes is closely related to the permafrost temperature, elevation, slope and degradation caused by climate warming [28]. The ground temperature and snow depth affect the thermokarst lake shoreline in subarctic peat plateau landscapes [29]. Moreover, the vegetation during growing season is not only an indictor of CO_2 and CH_4 exchange rates, but also can affect the rate of thermokarst lake formation and drainage [30]. In Yakutia, East Siberia, the co-acting factors of temperature rise and higher precipitation rates promote thermokarst lake enlargement and dynamics [31]. Lu et al. chose spot and area densities, frozen ground types, the mean annual ground temperature, vegetation coverage, soil type and terrain slope to evaluate the susceptibility of thermokarst lakes along the Qinghai–Tibet railway [32]. Based on the elevation, slope, aspect, topographic position index (TPI), height above the nearest drainage (HAND), stream power index (SPI), topographic wetness index (TWI), mean annual ground temperature (MAGT), active layer thickness (ALT), rainfall, Normalized Difference Vegetation Index (NDVI) [33], and fine soil content and machine learning models, the authors conducted thermokarst lake susceptibility assessments in permafrost landscapes of the Arctic [34–36].

In summary, although previous research on the extraction of thermokarst lakes in the Qinghai–Tibet Plateau has made significant progress, the current study relies on existing remote sensing data and literature sources. In areas that are difficult to access and where data collection is challenging, some data sources come from different years, which may introduce certain errors. Although the overall trend remains unaffected, this still poses a challenge regarding precision. In such a unique region as the Qinghai–Tibet Plateau, the complexity of factors such as topography and climate makes it difficult to precisely delineate the boundaries and forms of thermokarst lakes. Furthermore, existing technologies are limited in their ability to capture spatial features at local scales, making it difficult to accurately identify thermokarst lakes. While the study considered multiple environmental factors, a deeper understanding of the mechanisms behind thermokarst lake formation is still required, especially in terms of how human activities, climate change, and other factors may further exacerbate the expansion of these lakes and their impact on the ecological environment.

We interpreted the number and area of thermokarst lakes on the QTP based on remote sensing, combined with field investigation. The objectives of this research are to (1) interpret the thermokarst lakes present in 2020 to determine the thermokarst lake density based on remote sensing, (2) determine the assessment index system, (3) confirm the weight of the assessing factors, and (4) perform thermokarst lake susceptibility assessments based on a comprehensive evaluation method.

The thermokarst lake susceptibility evaluation index system proposed in this study provides an effective tool for natural disaster early warning on the Qinghai–Tibet Plateau and in similar high-altitude regions. By identifying high-susceptibility areas and clarifying the main factors influencing the distribution of thermokarst lakes, this research offers a scientific basis for ecological protection, land-use planning, and infrastructure development on the Qinghai–Tibet Plateau. In particular, in high-risk areas, rational engineering design and research not only deepen our understanding of thermokarst lake dynamics, but also provide predictive insights into the expansion of thermokarst lakes under future climate change scenarios. This holds long-term significance for regional ecological restoration and climate adaptation policy formulation. Environmental protection measures can effectively reduce the negative impacts of thermokarst lakes on ecosystems and socio-economic activities.

2. Materials and Methods

2.1. Study Area

The QTP is located in western China, spanning Xizang, Qinghai, and parts of Xinjiang, Gansu, Sichuan, Yunnan and other provinces and regions. Known as the Third Pole of the Earth, it comprises an area of about 256×10^4 km^2, with a latitudinal gradient of 26–40° N and a longitude of 73–104° E [37] (Figure 1). The water resources of the Qinghai–Tibet Plateau exist in various forms such as rivers, lakes, glaciers, and groundwater, with river runoff as the main source. The QTP is home to numerous glaciers and is the birthplace of over ten major Asian rivers, such as the Yellow River, Yangtze River, Yarlung Zangbo River, Nujiang River, etc., and thus is widely acknowledged as the Asian Water Tower [38,39]. A low temperature, strong radiation and a lower cumulative temperature are typical climate characteristics of the QTP. These special climatic features make the QTP the largest expanse of high-altitude permafrost worldwide. The permafrost in this area is characterized by a higher mean annual ground temperature, a considerable ground ice content, an elevated soil carbon content, and remarkable climate sensitivity. Therefore, the QTP is recognized as an amplifier and regulator of climate change in the Northern Hemisphere [37]. Moreover, lakes are scattered on the QTP, and perform a dazzling array of vital ecological functions.

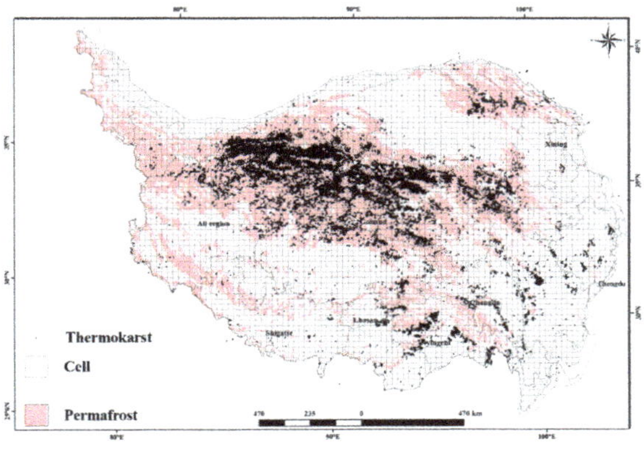

Figure 1. Assessment units and LD.

2.2. Data Preparation

Lake point density (LD) reflects the concentration and spread of thermokarst lakes within the study area. A higher density suggests an increased occurrence of thermokarst lakes, which is critical for identifying high-susceptibility regions.

Soil moisture (SM) plays a crucial role in the formation of thermokarst lakes, as it directly affects the thawing of permafrost and the hydrological processes in the region.

Slope is a critical factor in the stability of permafrost and the formation of thermokarst lakes. Areas with low slope gradients tend to accumulate water more easily, which may lead to the formation of thermokarst lakes.

The NDVI is used to assess vegetation cover and productivity, as vegetation can serve as an indicator of the environmental conditions of the region; in the context of thermokarst lakes, areas with lush vegetation may indicate more stable conditions for lake formation, while snow cover influences the thermal insulation of the ground and can affect the rate of permafrost thawing.

The snow cover area (SCA) can affect the thermal insulation of the ground and the rate of permafrost thawing. Snow accumulation can protect the underlying permafrost from extreme winter temperatures, influencing the stability of thermokarst lakes in the subsequent thawing season.

The mean annual ground temperature (MAGT) is a direct indicator of the thermal conditions of the permafrost, influencing its stability.

The mean annual precipitation (MAP) is a key factor in determining the water availability in a region. Areas with higher precipitation are more likely to experience water accumulation, which can facilitate the formation and expansion of thermokarst lakes.

The stability of permafrost (PS) is a central factor in the formation of thermokarst lakes. Regions with unstable or thawing permafrost are more prone to the formation of thermokarst lakes due to the collapse of the ground structure. This indicator reflects the susceptibility of permafrost to thawing and subsequent lake formation.

The data on thermokarst lakes were acquired by computer classification and the visual interpretation of Landsat images. The Landsat image data selected were the Level 1 products of Landsat-5 TM and Landsat-8 OLI (Level 1 Precision Terrain Corrected). The data were calculated using ENVI 5.3 software for average altitude, radiometric calibration, and atmospheric correction, completing the preliminary preparation work. Slope was calculated from the DEM. The NDVI was calculated as the maximum of the annual NDVI in 2020 based on Moderate-Resolution Imaging Spectroradiometer (MODIS) data. SM and MAP data are provided by the National Tibetan Plateau/Third Pole Environment Data

Center (http://data.tpdc.ac.cn). PS, MAGT and SCA data are from "A Big Data Platform for Three Poles" [40].

2.3. Assessment Index System

Assessment index systems are the foundation of modeling and mapping. The construction of an evaluation index system should follow these principles: systematicness, representativeness, dynamicity, quantification, conciseness, comprehensiveness, etc. In this study, factors were selected considering the characteristics of thermokarst lake development and recent research achievements [41–45]. The eight evaluation indicators included LD, SM, slope, NDVI, SCA, MAGT, MAP and PS. Different evaluation indicators have different dimensions and units, which can affect the results of data analysis. In order to eliminate the dimensional impacts between indicators, data normalization (Equation (1)) was performed to address the comparability between evaluation indicators. According to the relationship between thermokarst lake distribution and slope, it was found that lakes are mainly distributed within a slope range of less than 30°. So, in this normalization calculation, we assigned a value greater than 30° to 0. The PS types include very stable, stable, semi-stable, transitional and unstable. The normalized values are 0.1, 0.3, 0.5, 0.7, and 1, respectively.

$$x' = \frac{x - min(x)}{max(x) - min(x)} \tag{1}$$

where x is the original value of an evaluation index, $min\ (x)$ is the minimum value of the index in all samples, $max\ (x)$ is the maximum value of the index in all samples, and x' is the normalized index value $\in$ (0,1).

After normalizing all evaluation indicators, we obtained the following figures (Figures S1–S7 are available in the Supplementary document).

The density of thermokarst lakes represents the degree of the development of thermokarst lakes in the study area. The higher the density of points, the greater the likelihood of thermokarst lakes occurring. Slope not only determines the runoff of surface water, the supply and discharge of groundwater, and vegetation coverage, but also affects its stress distribution. Moreover, slope is also a controlling factor for the stability of thermokarst lakes. Permafrost-related data, such as PS, SCA and MAGT, are major indicators of the permafrost thermal state and affect the development of thermokarst lakes. Vegetation plays an essential role in slope protection and the prevention of soil erosion, and has a certain impact on the evolution and stability of slopes. Meteorological factors, such as MAP, are the main source of water supply for thermokarst lakes, being supplied through surface runoff and water above the frozen layer [19]. The soil composition includes clay, silt and sand on the QTP. SM quantifies hydrological processes and helps to accommodate excess ice [41], which facilitates the formation of thermokarst lakes.

The weight of evaluation indicators refers to the relative importance of a certain factor in the overall evaluation. The higher the weight, the more important this factor is. For the determination of weights, this paper introduces the relevant literature [32] and adopts the expert scoring method, as shown in Table 1.

Table 1. Susceptibility assessment indexes and weight.

Indexes	Weight	Indexes	Weight
LD	0.3	SCA	0.08
SM	0.05	MAGT	0.11
Slope	0.02	MAP	0.25
NDVI	0.04	PS	0.15

2.4. Assessment Unit

Due to the large area of the research area, it has been divided into 2.5 km × 2.5 km units, with a total of 3397 units in permafrost regions. The unit assessment method can take into account the differences in the index across the area by making the unit an evaluation unit.

2.5. Assessment Method

The comprehensive evaluation method (Equation (2)) transforms multiple indicators into one that can reflect the comprehensive situation for analysis and assessment, and is used for the overall evaluation of things that are restricted by multiple factors [46]. This method is objective, scientific, credible and can combine qualitative and quantitative indicators [47]. It has great advantages in coping with problems that are uncertain and difficult to quantify. Based on the use of the comprehensive evaluation method and expert scoring method, the establishment of the thermokarst lake susceptibility evaluation system was evaluated.

$$S = \sum_{i=1}^{n} W_i \times x_i' \quad (2)$$

where S is the comprehensive score, W_i is the weight of each evaluation indicator, and x_i' is the normalized index value within each unit. A higher calculated susceptibility index S indicates a higher potential for thermokarst lake occurrence.

Using the weights calculated by the expert score theory combined with comprehensive evaluation, the final comprehensive susceptibility index S was calculated. The study applied ArcGIS 10.5 software to overlay and analyze the susceptibility of thermokarst lakes with related evaluation indices, and based on change-point, classified the degree of susceptibility into Very low (0.1~0.22), Low (0.22~0.28), Moderate (0.28~0.37), and High (0.37~0.59) (Table 2).

Table 2. The degree of thermokarst lake susceptibility assessment in the QTP.

Classes	Very Low	Low	Moderate	High
S	0.1~0.22	0.22~0.28	0.28~0.37	0.37~0.59

The term "most susceptible" is used to describe the areas, groups, or individuals that have the highest likelihood of being affected by a hazard or issue, based on the analysis and determination using the natural breaks method. These areas, groups, or individuals are at the highest risk and require the most attention and mitigation efforts.

2.6. Evaluation Index Importance Based on Random Forest

To better understand the relative importance of the eight selected evaluation indicators (lake point density, soil moisture, slope, NDVI, snow cover area, MAGT, mean annual precipitation, and permafrost stability) in the formation of thermokarst lakes, we employed the Random Forest (RF) algorithm. This machine learning technique was used to assess the contribution of each factor by ranking their importance, based on their predictive power, in determining thermokarst lake susceptibility.

The Random Forest model was trained using the selected factors as input variables and the susceptibility assessment indexes of thermokarst lakes as the target variable. The importance of each evaluation indicator was then calculated using the "Mean Decrease Impurity" (MDI) method, which evaluates the contribution of each variable to reducing the impurity in the decision trees. A higher MDI score indicates that a given factor has a greater impact on the model's predictive accuracy, thus being more influential in determining the spatial distribution of thermokarst lakes.

The Random Forest analysis revealed the relative importance of each factor, highlighting which environmental variables play a dominant role in thermokarst lake formation. The calculation formula is as follows:

$$MDI_f = \frac{1}{T} \sum_{s \in \text{splits}} \Delta I_s^t \quad (3)$$

where MDI_f is the mean decrease impurity of feature f; T is the total number of decision trees in the forest; ΔI_s^t is the decrease in impurity for split s in tree t; $\sum_{s \in \text{splits}}$ represents the summing of the impurity decreases across all splits in tree t where feature f is used.

3. Results

The area of permafrost in these statistics is 105.06×10^4 km^2.

We performed area statistics of the thermokarst lake susceptibility in the study area according to the fan classification criteria in Table 2, and the results are shown in Table 3.

Table 3. Thermokarst lake susceptibility assessment result for the QTP.

Classes	Area (km^2)	Area Covered	Main Distribution Region
Very low	24.25×10^4	23.08%	Central Bayingolin Mongolian Autonomous Prefecture
Low	33.87×10^4	32.24%	Southern Bayingolin Mongolian Autonomous Prefecture, Central Nagqu, Central Golmud
Moderate	26.96×10^4	25.66%	Central Ali, Eastern Yushu
High	19.98×10^4	19.02%	Southwest of Qinghai Province, Northeast of Xizang, Hoh Xil region

3.1. High Susceptibility Area

Most thermokarst lakes are found in high-susceptibility areas, characterized by relatively flat terrain. These areas, located primarily in the southwest of Qinghai Province, northeast Tibet (Xizang), and the Hoh Xil region, cover approximately 19.98×10^4 km^2, or 19.02% of the study area (Table 3). The permafrost here is mainly semi-stable to transitional, prone to thawing under certain conditions but not as unstable as other regions. The climate and geological characteristics of this region contribute to its susceptibility. The MAGT ranges from −0.5 °C to −3.0 °C, with permafrost thickness between 30 m and 110 m. The average annual temperature is between −4.0 °C and −6.5 °C, which, combined with precipitation levels of 200–400 mm, creates favorable conditions for thermokarst lake formation due to ground thawing processes. Human activities, such as highway and railway construction, have further altered the region's landscape, triggering environmental changes that promote thermokarst lake formation. These disturbances increase the region's vulnerability, exacerbating risks to both local ecosystems and infrastructure [48]. Given these factors, effective monitoring and management are essential to mitigate potential environmental and infrastructural impacts.

3.2. Moderate Susceptibility Area

This area includes regions with small slopes, semi-stable permafrost, and low ground temperature. The characteristics of these regions, including relatively stable permafrost conditions and lower temperature fluctuations, result in a scattered distribution of thermokarst lakes across the assessment area. The distribution pattern of thermokarst lakes in this type of region is largely influenced by both the density of existing thermokarst lakes and the stability of the underlying permafrost. In these areas, permafrost is generally in a semi-stable state, meaning it is prone to thawing under certain conditions but does not exhibit the extreme instability seen in other more vulnerable regions. The influence of thermokarst lake density is particularly important in these areas. The presence of numerous smaller lakes can contribute to further destabilization of the surrounding land, even if the lakes themselves are not as expansive as those found in areas with higher susceptibility. However, due to the relatively low slope of the terrain and lower ground temperature, the potential for large-scale thermal erosion and collapse of the permafrost is somewhat reduced compared to more vulnerable areas.

In terms of the spatial extent, this region occupies approximately 26.96×10^4 km^2, accounting for about 25.66% of the total study area (Table 3). The moderate susceptibility of this region indicates that, while thermokarst lake formation is still a significant concern, it is not as imminent or widespread as in the high susceptibility areas. Nevertheless, these regions are still susceptible to the gradual expansion of thermokarst lakes over time, especially if there are changes in environmental conditions such as rising temperatures or increased human activities.

The presence of moderate slope gradients, coupled with the semi-stable nature of the permafrost and relatively lower ground temperatures, suggests that these regions could experience slower but steady changes in landscape patterns, with the gradual formation of thermokarst lakes over time. Additionally, the relative stability of the permafrost in these areas may allow for the accumulation of moisture and the development of wetland environments, further promoting the growth of thermokarst lakes under the right climatic and hydrological conditions.

3.3. Low Susceptibility Area

The low susceptibility area is scattered around the moderate susceptibility region and represents a more stable environment. In these areas, permafrost remains stable, and the vegetation coverage is relatively low, primarily consisting of alpine meadows. These meadows play an important role in the region's thermal dynamics, as their dense vegetation helps form an insulating layer during the winter months. This layer prevents heat loss from the soil, reducing the rate at which the permafrost thaws. As a result, the stability of the permafrost is maintained, and the risk of thermokarst lake formation is lower. However, the low vegetation coverage and thin soil layers mean that this region is still sensitive to environmental changes. Although thermokarst lakes are less likely to form here compared to other areas, any significant alteration in climate conditions or human activities could destabilize the permafrost, leading to potential landscape changes and localized ecological disturbances.

3.4. Very Low Susceptibility Area

Figure 2 shows that the "very low susceptibility area" is characterized by a dispersed distribution. The formation of thermokarst lakes in this region is the result of a complex interplay of multiple factors. Precipitation and SM are key determinants in maintaining water storage in thermokarst lakes. In this area, the annual mean precipitation is relatively low and occurs mainly during the summer and winter months, which limits the amount of water available for lake formation. Additionally, the soil is classified as alpine desert soil, with a thin layer, abundant gravel, and minimal fine particles, leading to low SM. These soil characteristics reduce the capacity for water retention, further limiting the potential for thermokarst lake development. As a result, this area experiences minimal thermokarst lake formation, as the conditions do not support significant ground thawing or water accumulation.

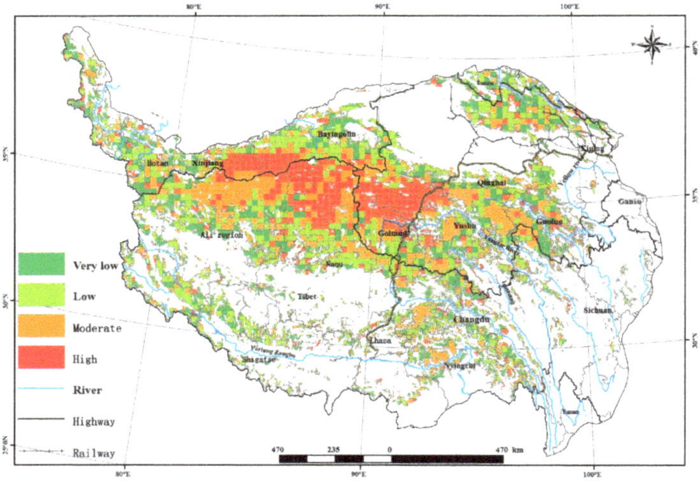

Figure 2. Zoning Map of the Susceptibility of thermokarst Lake in Permafrost Regions.

3.5. Assessment of Evaluation Index Importance Based on Random Forest

The MDI scores for each feature can be obtained after running the code with python (Table 4).

Table 4. Evaluation Indicator MDI Score (Importance).

Evaluation Indicator	MDI Score	Evaluation Indicator	MDI Score
LD	0.18	SCA	0.10
SM	0.16	MAGT	0.15
Slope	0.14	MAP	0.08
NDVI	0.12	PS	0.07

In this set, the importance is more evenly distributed across the factors, indicating that no single factor overwhelmingly dominates. LD, AM, and MAGT are the most influential factors, with PS and MAP contributing less.

4. Discussion

4.1. Key Drivers and Comparative Analysis of Thermokarst Lake Susceptibility

The susceptibility of thermokarst lakes is based on geological environmental conditions, combined with meteorological factors and triggering factors, to predict the likelihood of their occurrence in a certain area. By integrating multiple conditioning factors, the weight of each factor is determined by expert experience, and finally, the spatial distribution probability of thermokarst lakes in the region is obtained.

In terms of the environmental factors, our study confirms that permafrost stability PS, slope, and precipitation are key drivers of thermokarst lake formation. This finding resonates with earlier works that have highlighted the importance of ground temperature and permafrost thawing in influencing thermokarst dynamics. Furthermore, studies have also recognized the role of human activities in accelerating thermokarst lake formation. Our findings align with these studies, as infrastructure development—such as highways and railways—has been shown to exacerbate the thawing of permafrost and the expansion of thermokarst lakes. These anthropogenic activities disrupt the natural balance, promoting conditions that favor lake formation, particularly in regions with moderate PS.

At the same time, changes in the soil composition and ice content in different regions of the Qinghai–Tibet Plateau can also impact the formation of thermokarst lakes. In some areas, soils may be rich in clay or organic matter, which can affect the soil's water retention capacity and thermal conductivity. When soil moisture is higher, it may be more susceptible to temperature variations, thereby promoting permafrost thawing and the formation of thermokarst lakes [49]. In regions with a high ice content, as the climate warms, the melting of underground ice layers is more likely to result in accumulated meltwater, which can form thermokarst lakes.

Previous studies have used machine learning models and training samples to assess the thermokarst lake susceptibility in the Qinghai–Tibet Engineering Corridor [41]. It was confirmed that more than 90% of the thermokarst lake points are in high susceptibility areas [23]. Niu et al. calculated that 20.79% of the Qinghai–Tibet Engineering Corridor is in a very high susceptibility area. Because the Qinghai–Tibet Engineering Corridor is a part of the QTP, the conclusions drawn by the above scholars are highly consistent with ours [50]. Scholars use different evaluation indicators, resulting in varying susceptibility outcomes in some regions; for example, Yin et al. [43] used MAAT, landform, vegetation, surficial geology, MAGT, the ground ice condition and lake number, while Wang et al. [34] used elevation, slope, aspect, TPI, HAND, SPI, TWI, MAGT, ALT, rainfall, NDVI and fine soil content.

The Random Forest analysis shows that the importance of the factors influencing thermokarst lake formation is relatively evenly spread, meaning no single factor dominates the process. The most important factors are LD, SM, and MAGT. These factors directly affect the distribution and formation of thermokarst lakes by influencing the thawing of permafrost and water availability. In contrast, PS and MAP have a lower contribution. While they are still important, they play a less direct role compared to temperature and moisture. This suggests that thermokarst lake formation is shaped by multiple interacting factors, with temperature and moisture being the key drivers. Future studies should focus on how these factors interact and how they might change under climate change or human influence.

4.2. The Evaluation Index System for Thermokarst Lake Susceptibility Provides an Effective Tool for Natural Disaster Assessment in the Qinghai–Tibet Plateau

Establishing an effective evaluation model and conducting susceptibility assessments play a crucial role in improving the efficiency and accuracy of regional disaster prediction and early warning systems. In the context of the Qinghai–Tibet Plateau (QTP), where climate change and permafrost degradation significantly influence the landscape and infrastructure, such assessments are essential for proactive management and mitigation. However, the challenge lies in developing an evaluation index system that is both scientifically robust and tailored to the unique environmental and climatic conditions of the region. The scientific process of determining the most relevant and impactful factors for thermokarst lake formation, and integrating them into a usable index, is not straightforward. Establishing an index system that is aligned with regional practical needs requires the careful selection of parameters that are both locally applicable and scientifically sound.

In the absence of sufficient, region-specific measured data, this study relies on previous research findings and the available data, ensuring that the selection of evaluation indicators aligns with the known environmental dynamics of the QTP. This approach, while effective, does present some limitations, particularly with regard to the temporal consistency of the data. The issue of using data from different years may introduce some degree of error, particularly for parameters that are sensitive to temporal changes. However, it is important to note that such discrepancies in data are unlikely to have a significant impact on the overall trend or the spatial distribution of susceptibility. The general conclusions drawn from this approach still provide valuable insights into the regions that are most at risk of thermokarst lake formation and the environmental factors that contribute to it.

4.3. Uncertainties and Future Perspectives

Despite the significant progress made in mapping and assessing the susceptibility of thermokarst lakes on the QTP, several uncertainties that affect the overall accuracy and reliability of the results remain. These uncertainties arise from a variety of sources, including data limitations, methodological constraints, and environmental variability. One of the primary challenges in this study lies in the availability and quality of the data used for susceptibility evaluation. Due to the remoteness and inaccessibility of many areas in the QTP, there is a lack of sufficient ground truth data, which limits our ability to accurately assess the current state of thermokarst lakes and other key environmental parameters. In particular, the accurate delineation of thermokarst lake boundaries and their dynamic changes over time remains a major difficulty.

One of the primary challenges in this study is the availability and quality of the data used for susceptibility evaluation. The remoteness and inaccessibility of many areas in the QTP have resulted in a lack of sufficient ground truth data, limiting our ability to accurately assess the current state of thermokarst lakes and other key environmental parameters. In particular, the accurate delineation of thermokarst lake boundaries and their dynamic changes over time remains a major difficulty. Despite advances in remote sensing technologies and satellite imagery, they still struggle to capture the fine spatial details necessary for precise mapping, particularly in regions with a complex terrain and high altitude. Another source of uncertainty stems from the temporal variation in key environmental factors such as temperature, precipitation, and PS. Given that the data used in this study may not be from the same year or season, there could be discrepancies in how these factors are represented in the analysis. While these discrepancies do not significantly alter the overall trend or large-scale patterns of susceptibility, they affect the precision of specific results and their predictive power for short-term changes. Additionally, the variability in seasonal patterns—especially the timing and intensity of precipitation or snowmelt—compounds the challenges associated with understanding the true dynamics of thermokarst lakes and their seasonal fluctuations.

Current remote sensing technologies and satellite imagery have advanced, but they still struggle to capture the fine spatial details required for precise mapping, especially in regions with complex terrain and high altitude. Moreover, the temporal variation in key factors such as temperature, precipitation, and PS adds another layer of uncertainty [51]. Given that the data used in this study may not be from the same year or season, there could be discrepancies in how these factors are represented in the analysis. While this may not significantly alter the overall trend or the large-scale patterns of susceptibility, it does affect the precision of specific results and their predictive power for short-term changes. The variability in seasonal patterns, especially the timing and intensity of precipitation or snowmelt, also introduces challenges in understanding the true dynamics of thermokarst lakes and their seasonal fluctuations.

However, due to the difficulty of accurately defining and identifying thermokarst lakes, detailed information on thermokarst lakes in the QTP region is still scarce [52]. In the permafrost region, the ground ice content is often used to define the spatial extent of thermokarst lake development [4,12]. The ground ice content layer has an extremely broad regionalization; unfortunately, with current technology, finer spatial features cannot be captured at local scales [10,41].

Looking to the future, addressing these uncertainties will require advancements in several areas. First, further field investigations and long-term monitoring are essential to gather more reliable data on the current state of and changes in thermokarst lakes. The increased availability of high-resolution satellite imagery and aerial drones can help improve the spatial accuracy of mapping efforts and provide real-time monitoring capabilities. Additionally, more synchronized and consistent data, covering the same time periods, should be integrated into the analysis to reduce temporal discrepancies. Advances in machine learning and artificial intelligence could also enhance our ability to model and predict thermokarst lake dynamics more accurately. These technologies, coupled with better data integration, may enable researchers to develop more robust models that account for the complex interactions between environmental factors and human activities. The use of big data and high-performance computing could facilitate the development of dynamic, real-time prediction systems for thermokarst lake evolution, allowing for more effective early warning systems and more targeted conservation strategies.

5. Conclusions

Unsupervised classification and interactive visualization interpretation were adopted to interpret the spatial distribution of thermokarst lakes over the QTP.

(1) The results revealed that 19.02% of the total permafrost area is highly susceptible to thermokarst lake formation. The high-susceptibility zones are primarily located in the southwest of Qinghai Province, northeast Tibet (Xizang), and the Hoh Xil region.
(2) This study identified PS and slope as the primary factors influencing the spatial distribution of thermokarst lakes. These findings emphasize the need to pay attention to the impacts of thermokarst lakes on the stability of infrastructure in these regions.
(3) This research contributes to a deeper understanding of thermokarst lake development and the environmental factors involved. Additionally, it provides valuable insights into the long-term dynamics of thermokarst lakes in permafrost regions, offering a foundation for future studies and environmental management strategies.

In conclusion, while this study has provided valuable insights into thermokarst lake susceptibility, several limitations should be acknowledged. Firstly, the methodologies employed and the expert scoring method used to determine the weights in this article are highly subjective and lack objectivity. Secondly, in the absence of data, the selected indicators are few and provide insufficient representativeness.

Supplementary Materials: The following supporting information can be downloaded at: https://www.mdpi.com/article/10.3390/su17041464/s1.

Author Contributions: Conceptualization, L.L., W.N. and F.N.; methodology, L.L. and X.L.; data curation, X.L.; writing—original draft preparation, L.L. and Y.Z.; writing—review and editing, W.N. and Y.Z.; funding acquisition, L.L. All authors have read and agreed to the published version of the manuscript.

Funding: This research was funded by the Second Tibetan Plateau Scientific Expedition and Research Program (STEP), grant number 2019QZKK0905; the Science and Technology Project of Henan Province grant number 222102320451 and 232102321016; the National Natural Science Foundation of China (Grant No. 42261144749, 42377158); College Students' Innovative Entrepreneurial Training Plan Program (S202410710327; S202410710331); International Science and Technology Cooperation Program of Shaanxi Province (Grant No. 2024GH-ZDXM-24). Shaanxi Province Agricultural science and technology 114 public welfare platform to serve rural revitalization practical technical training (Grant No. 2024NC-XCZX-06).

Institutional Review Board Statement: Not applicable.

Informed Consent Statement: Not applicable.

Data Availability Statement: The original contributions presented in this study are included in the article. Further inquiries can be directed to the corresponding authors.

Acknowledgments: We sincerely thank the editor and all the reviewers for their valuable comments and feedback. We also thank the National Tibetan Plateau Data Center for providing the additional dataset.

Conflicts of Interest: The authors declare no conflicts of interest.

References

1. Farquharson, L.M.; Romanovsky, V.E.; Cable, W.L.; Walker, D.A.; Kokelj, S.V.; Nicolsky, D. Climate change drives widespread and rapid thermokarst development in very cold permafrost in the Canadian High Arctic. *Geophys. Res. Lett.* **2019**, *46*, 6681–6689. [CrossRef]
2. Pandey, P.; Ali, S.N.; Sharma, V.; Champati Ray, P.K. Focus on Thermokarst Lakes in Indian Himalaya: Inception and implication under warming climate. *J. Clim. Change* **2020**, *6*, 59–69. [CrossRef]
3. Cusicanqui, D.; Bodin, X.; Duvillard, P.A.; Schoeneich, P.; Revil, A.; Assier, A.; Berthet, J.; Peyron, M.; Roudnitska, S.; Rabatel, A. Glacier, permafrost and thermokarst interactions in Alpine terrain: Insights from seven decades of reconstructed dynamics of the Chauvet glacial and periglacial system (Southern French Alps). *Earth Surf. Process. Landf.* **2023**, *48*, 2595–2612. [CrossRef]
4. Grosse, G.; Jones, B.M.; Arp, C.D. *Thermokarst Lakes, Drainage, and Drained Basins*; Elsevier: Amsterdam, The Netherlands, 2013.
5. Hu, B.; Wu, Y.; Zhang, X.; Yang, B.; Chen, J.; Li, H.; Chen, X.; Chen, Z. Monitoring the Thaw Slump-Derived Thermokarst in the Qinghai-Tibet Plateau Using Satellite SAR Interferometry. *J. Sens.* **2019**, *2019*, 1698432. [CrossRef]
6. Yang, Y.; Rogers, B.M.; Fiske, G.; Watts, J.; Potter, S.; Windholz, T.; Mullen, A.; Nitze, I.; Natali, S.M. Mapping retrogressive thaw slumps using deep neural networks. *Remote Sens. Environ.* **2023**, *288*, 113495. [CrossRef]
7. Lewkowicz, A.G.; Way, R.G. Extremes of summer climate trigger thousands of thermokarst landslides in a High Arctic environment. *Nat. Commun.* **2019**, *10*, 1329. [CrossRef]
8. Bertran, P.; Andrieux, E.; Bateman, M.; Font, M.; Manchuel, K.; Sicilia, D. Response to the discussion by Van Vliet-Lanoë et al. of the paper 'Features caused by ground ice growth and decay in Late Pleistocene fluvial deposits, Paris Basin, France'(Bertran et al., 2018, Geomorphology 310, 84–101). *Geomorphology* **2019**, *327*, 629–633. [CrossRef]
9. Hjort, J.; Streletskiy, D.; Doré, G.; Wu, Q.; Bjella, K.; Luoto, M. Impacts of permafrost degradation on infrastructure. *Nat. Rev. Earth Environ.* **2022**, *3*, 24–38. [CrossRef]
10. Nitze, I.; Heidler, K.; Barth, S.; Grosse, G. Developing and testing a deep learning approach for mapping retrogressive thaw slumps. *Remote Sens.* **2021**, *13*, 4294. [CrossRef]
11. Șerban, R.-D.; Jin, H.; Șerban, M.; Luo, D.; Wang, Q.; Jin, X.; Ma, Q. Mapping thermokarst lakes and ponds across permafrost landscapes in the Headwater Area of Yellow River on northeastern Qinghai-Tibet Plateau. *Int. J. Remote Sens.* **2020**, *41*, 7042–7067. [CrossRef]
12. Luo, J.; Niu, F.; Lin, Z.; Liu, M.; Yin, G.; Gao, Z. Abrupt increase in thermokarst lakes on the central Tibetan Plateau over the last 50 years. *Catena* **2022**, *217*, 106497. [CrossRef]
13. Veremeeva, A.; Nitze, I.; Günther, F.; Grosse, G.; Rivkina, E. Geomorphological and climatic drivers of thermokarst lake area increase trend (1999–2018) in the Kolyma lowland yedoma region, north-eastern Siberia. *Remote Sens.* **2021**, *13*, 178. [CrossRef]
14. Heslop, J.K.; Anthony, K.W.; Winkel, M.; Sepulveda-Jauregui, A.; Martinez-Cruz, K.; Bondurant, A.; Grosse, G.; Liebner, S. A synthesis of methane dynamics in thermokarst lake environments. *Earth-Sci. Rev.* **2020**, *210*, 103365. [CrossRef]
15. Rodenhizer, H.; Belshe, F.; Celis, G.; Ledman, J.; Mauritz, M.; Goetz, S.; Sankey, T.; Schuur, E.A. Abrupt permafrost thaw accelerates carbon dioxide and methane release at a tussock tundra site. *Arct. Antarct. Alp. Res.* **2022**, *54*, 443–464. [CrossRef]
16. Kuhn, M.A.; Schmidt, M.; Heffernan, L.; Stührenberg, J.; Knorr, K.H.; Estop-Aragonés, C.; Broder, T.; Gonzalez Moguel, R.; Douglas, P.M.; Olefeldt, D. High ebullitive, millennial-aged greenhouse gas emissions from thermokarst expansion of peatland lakes in boreal western Canada. *Limnol. Oceanogr.* **2023**, *68*, 498–513. [CrossRef]
17. Manasypov, R.M.; Lim, A.G.; Krickov, I.; Shirokova, L.S.; Shevchenko, V.P.; Aliev, R.A.; Karlsson, J.; Pokrovsky, O.S. Carbon storage and burial in thermokarst lakes of permafrost peatlands. *Biogeochemistry* **2022**, *159*, 69–86. [CrossRef]
18. Wang, L.; Du, Z.; Wei, Z.; Xu, Q.; Feng, Y.; Lin, P.; Lin, J.; Chen, S.; Qiao, Y.; Shi, J. High methane emissions from thermokarst lakes on the Tibetan Plateau are largely attributed to ebullition fluxes. *Sci. Total Environ.* **2021**, *801*, 149692. [CrossRef]
19. Gao, Z.; Niu, F.; Wang, Y.; Wang, W.; Lin, Z.; Luo, J.; Lyu, M. Hydrological characteristics of thermokarst lake and its environmental effects on Qinghai-Tibet Plateau. *Adv. Water Sci.* **2022**, *33*, 174–184.
20. Wan, C.; Gibson, J.; Shen, S.; Yi, Y.; Yi, P.; Yu, Z. Using stable isotopes paired with tritium analysis to assess thermokarst lake water balances in the Source Area of the Yellow River, northeastern Qinghai-Tibet Plateau, China. *Sci. Total Environ.* **2019**, *689*, 1276–1292. [CrossRef]

21. Morgenstern, A.; Grosse, G.; Günther, F.; Fedorova, I.; Schirrmeister, L. Spatial analyses of thermokarst lakes and basins in Yedoma landscapes of the Lena Delta. *Cryosphere* **2011**, *5*, 849–867. [CrossRef]
22. Serikova, S.; Pokrovsky, O.; Laudon, H.; Krickov, I.; Lim, A.; Manasypov, R.; Karlsson, J. High carbon emissions from thermokarst lakes of Western Siberia. *Nat. Commun.* **2019**, *10*, 1552. [CrossRef] [PubMed]
23. Luo, J.; Niu, F.; Lin, Z.; Liu, M.; Yin, G. Thermokarst lake changes between 1969 and 2010 in the beilu river basin, Qinghai–tibet plateau, China. *Sci. Bull.* **2015**, *60*, 556–564. [CrossRef]
24. Chen, X.; Mu, C.; Jia, L.; Li, Z.; Fan, C.; Mu, M.; Peng, X.; Wu, X. High-resolution dataset of thermokarst lakes on the Qinghai-Tibetan Plateau. *Earth Syst. Sci. Data Discuss.* **2021**, *2021*, 1–23. [CrossRef]
25. Nitze, I.; Cooley, S.W.; Duguay, C.R.; Jones, B.M.; Grosse, G. The catastrophic thermokarst lake drainage events of 2018 in northwestern Alaska: Fast-forward into the future. *Cryosphere* **2020**, *14*, 4279–4297. [CrossRef]
26. Li, R.; Zhang, M.; Pei, W.; Melnikov, A.; Zhang, Z.; Li, G. Risk evaluation of thaw settlement using machine learning models for the Wudaoliang-Tuotuohe region, Qinghai-Tibet Plateau. *Catena* **2023**, *220*, 106700. [CrossRef]
27. Zhao, Z.; Huo, A.; Liu, Q.; Peng, J.; Elbeltagi, A.; Abuarab, M.E.-S.; Abu-Hashim, M.S.D. Spatiotemporal Variation in the Coupling Relationship between Human Activities and Soil Erosion—A Case Study in the Weihe River Basin. *Sustainability* **2023**, *15*, 10785. [CrossRef]
28. Liu, W.-H.; Zhou, G.-H.; Liu, H.-R.; Li, Q.-P.; Xie, C.-W.; Li, Q.; Zhao, J.-Y.; Zhang, Q. Large-scale characteristics of thermokarst lakes across the source area of the Yellow River on the Qinghai-Tibetan Plateau. *J. Mt. Sci.* **2023**, *20*, 1589–1604. [CrossRef]
29. Sannel, A.B.K. Ground temperature and snow depth variability within a subarctic peat plateau landscape. *Permafr. Periglac. Process.* **2020**, *31*, 255–263. [CrossRef]
30. Skeeter, J.; Christen, A.; Laforce, A.-A.; Humphreys, E.; Henry, G. Vegetation influence and environmental controls on greenhouse gas fluxes from a drained thermokarst lake in the western Canadian Arctic. *Biogeosciences* **2020**, *17*, 4421–4441. [CrossRef]
31. Czerniawska, J.; Chlachula, J. Climate-change induced permafrost degradation in Yakutia, East Siberia. *Arctic* **2020**, *73*, 509–528. [CrossRef]
32. Lu, J.; Cheng, H.; Niu, F.; Lin, Z.; Liu, H. Zoning evaluation on occurrence degree of thermokarst lake along Qinghai-Tibet Railway. *J. Catastrophol.* **2012**, *27*, 60–64.
33. Zhao, Y.; Huo, A.; Zhao, Z.; Liu, Q.; Zhao, X.; Huang, Y.; An, J. Novel Method for Evaluating Wetland Ecological Environment Quality Based on Coupled Remote Sensing Ecological Index and Landscape Pattern Indices: Case Study of Dianchi Lake Wetlands, China. *Sustainability* **2024**, *16*, 9979. [CrossRef]
34. Wang, R.; Guo, L.; Yang, Y.; Zheng, H.; Jia, H.; Diao, B.; Li, H.; Liu, J. Thermokarst lake susceptibility assessment using machine learning models in permafrost landscapes of the Arctic. *Sci. Total Environ.* **2023**, *900*, 165709. [CrossRef]
35. Baltiyeva, A.; Orynbassarova, E.; Zharaspaev, M.; Akhmetov, R. Studying sinkholes of the earth's surface involving radar satellite interferometry in terms of Zhezkazgan field, Kazakhstan. *Min. Miner. Depos.* **2023**, *17*, 61–74. [CrossRef]
36. Zapparov, M.; Kasenov, M.; Auelkhan, E.; Serikov, B.; Nurmukhanbet, S. Dangerous geological processes on Lake Kolsai and the impact of Anthropogenic load on the DGP. *Eng. J. Satbayev Univ.* **2022**, *144*, 46–51. [CrossRef]
37. Yao, T.; Thompson, L.G.; Mosbrugger, V.; Zhang, F.; Ma, Y.; Luo, T.; Xu, B.; Yang, X.; Joswiak, D.R.; Wang, W. Third pole environment (TPE). *Environ. Dev.* **2012**, *3*, 52–64. [CrossRef]
38. Xu, X.; Dong, L.; Zhao, Y.; Wang, Y. Effect of the Asian Water Tower over the Qinghai-Tibet Plateau and the characteristics of atmospheric water circulation. *Chin. Sci. Bull* **2019**, *64*, 2830–2841.
39. Ninglian, W.; Tandong, Y.; Baiqing, X.; An'an, C.; Weicai, W. Spatiotemporal pattern, trend, and influence of glacier change in Tibetan Plateau and surroundings under global warming. *Bull. Chin. Acad. Sci. (Chin. Version)* **2019**, *34*, 1220–1232.
40. Ran, Y.; Li, X.; Cheng, G.; Nan, Z.; Che, J.; Sheng, Y.; Wu, Q.; Jin, H.; Luo, D.; Tang, Z. Mapping the permafrost stability on the Tibetan Plateau for 2005–2015. *Sci. China Earth Sci.* **2021**, *64*, 62–79. [CrossRef]
41. Olefeldt, D.; Goswami, S.; Grosse, G.; Hayes, D.; Hugelius, G.; Kuhry, P.; McGuire, A.D.; Romanovsky, V.; Sannel, A.B.K.; Schuur, E. Circumpolar distribution and carbon storage of thermokarst landscapes. *Nat. Commun.* **2016**, *7*, 13043. [CrossRef]
42. Nitze, I.; Grosse, G.; Jones, B.M.; Romanovsky, V.E.; Boike, J. Remote sensing quantifies widespread abundance of permafrost region disturbances across the Arctic and Subarctic. *Nat. Commun.* **2018**, *9*, 5423. [CrossRef] [PubMed]
43. Yin, G.; Luo, J.; Niu, F.; Zhou, F.; Meng, X.; Lin, Z.; Liu, M. Spatial analyses and susceptibility modeling of thermokarst lakes in permafrost landscapes along the Qinghai–Tibet engineering corridor. *Remote Sens.* **2021**, *13*, 1974. [CrossRef]
44. Lara, M.J.; Chen, Y.; Jones, B.M. Recent warming reverses forty-year decline in catastrophic lake drainage and hastens gradual lake drainage across northern Alaska. *Environ. Res. Lett.* **2021**, *16*, 124019. [CrossRef]
45. Jones, B.M.; Grosse, G.; Farquharson, L.M.; Roy-Léveillée, P.; Veremeeva, A.; Kanevskiy, M.Z.; Gaglioti, B.V.; Breen, A.L.; Parsekian, A.D.; Ulrich, M. Lake and drained lake basin systems in lowland permafrost regions. *Nat. Rev. Earth Environ.* **2022**, *3*, 85–98. [CrossRef]
46. Zhang, F.; Hui, L.; Tao, S.; Yeerken, W. Assessment and mechanism analysis of the geo-integrated risk of Indochina Peninsula. *World Reg. Stud.* **2021**, *30*, 1151.

47. Luo, X.; Li, X.; Liu, W.; Wang, M.; Liu, C. Comprehensive evaluation of natural landscapes before and after earthquake based on GST-FAHP-GIS analytical method: A case study of Jiuzhaigou Nature Reserve, China. *Ecol. Indic.* **2023**, *154*, 110850. [CrossRef]
48. Du, R.; Peng, X.; Frauenfeld, O.W.; Sun, W.; Liang, B.; Chen, C.; Jin, H.; Zhao, Y. The role of peat on permafrost thaw based on field observations. *Catena* **2022**, *208*, 105772. [CrossRef]
49. Zhao, Z.; Huo, A.; Liu, Q.; Yang, L.; Luo, C.; Ahmed, A.; Elbeltagi, A. Assessment of urban inundation and prediction of combined flood disaster in the middle reaches of Yellow river basin under extreme precipitation. *J. Hydrol.* **2024**, *640*, 131707. [CrossRef]
50. Niu, F.; Lin, Z.; Liu, H.; Lu, J. Characteristics of thermokarst lakes and their influence on permafrost in Qinghai–Tibet Plateau. *Geomorphology* **2011**, *132*, 222–233. [CrossRef]
51. Huo, A.; Zhao, Z.; Luo, P.; Liu, Q.; Huang, Q.; Wang, Z.; Peng, J. Evolution of an arid social-ecosystem with different water utilization spanning 12,000 years. *J. Clean. Prod.* **2024**, *460*, 142548. [CrossRef]
52. Li, L.; Zhang, X.; Li, X.; Zhao, S.; Ni, W.; Yang, Z. Thermokarst lake changes over the past 40 years in the Qinghai–Tibet Plateau, China. *Front. Environ. Sci.* **2022**, *10*, 1051086. [CrossRef]

sustainability

MDPI

Article

Studies on Grass Germination and Growth on Post-Flotation Sediments

Grzegorz Żurek * and Danuta Martyniak

Plant Breeding and Acclimatization Institute, National Research Institute, Radzików, 05-870 Błonie, Poland; d.martyniak@ihar.edu.pl
* Correspondence: g.zurek@ihar.edu.pl

Abstract: The sediments remaining after copper flotation (here referred to as to flotation tailings) are generally characterized by nutrient deficiency and heavy metal enrichment, presenting significant obstacles to vegetation establishment and the development of sustainable ecosystems. This research aimed to evaluate the germination and growth performance of eight grass species on these tailings compared to reference conditions. Seed germination was assessed across flotation tailings, sand, and controlled laboratory environments. Initial establishment success was determined by measuring seedling height and root length three weeks after sowing in a glasshouse pot experiment. The findings revealed that *Bromus inermis*, *Bromus catharticus*, and *Elytrigia elongata* were capable of germinating and growing successfully in the challenging substrate, indicating their potential role in sustainable land management. Despite demonstrating excellent germination and seedling growth under test conditions, *Bromus tectorum* was deemed unsuitable due to its potential for invasive behavior, which could threaten the sustainability of revegetation efforts. Furthermore, species commonly employed for the revegetation of difficult terrains, including *Festuca arundinacea*, *Festuca rubra*, and *Lolium perenne*, proved entirely ineffective for the sustainable treatment of these flotation tailings.

Keywords: *Bromus tectorum*; *Bromus inermis*; *Bromus catharticus*; *Elytrigia elongata*; reclamation; post-mining heap

1. Introduction

Industry and intensive agriculture contribute significantly to the depletion of soil resources. These changes are often irreversible. Human-induced changes differ fundamentally from natural changes and have more extensive effects than anticipated or intended [1]. The primary drivers of land degradation include human activities such as deforestation, overgrazing, improper agricultural practices, and industrial emissions. These activities lead to soil erosion, loss of soil fertility, and consequently, biodiversity loss [2,3]. Degraded areas, for example, in Poland are particularly characterized by post-industrial sites, abandoned mines, un-reclaimed mining heaps, and urban regions suffering from socio-economic decline [4–6].

Heavy industries, especially in the Upper Silesian Industrial Region, have resulted in the widespread contamination of land with harmful substances and physical degradation due to mining and smelting activities [7,8]. Urban areas face degradation due to socio-economic decline, inadequate infrastructure, and high concentrations of negative social phenomena [9]. Despite the amount of scientific research and practical activities carried out to date, there are still issues that require urgent solution. One such issue is the growing amount of un-reclaimed remains of the mining industry, which are a clear example of unsustainable mining practices [10,11]. Due to the specificity of different mining processes, these remains are very different and, as such, require separate, specific approaches to their recultivation [12]. One of the many specific types of mining remains are flotation tailings of copper sulfide ores. The ore is of sedimentary origin and the tailing contains mainly quartz, dolomite, clay minerals, traces of sulfides, and some accessory minerals, as well as heavy metals. The waste handling and disposal system includes traditional 'slurry' transport combined with a hydraulic placement in an impoundment, which creates long-term environmental liabilities [12,13].

The problem of wind erosion from flotation tailings disposal sites poses significant environmental and health risks, particularly due to the high sand and silt content of these tailings [13,14]. After drying, they are very volatile due to their high content of sand and silt fractions. The dust from flotation tailing can be carried by the wind to surrounding areas, depositing heavy metals and other harmful substances onto soil and water bodies [15]. This can degrade air quality, harm vegetation, and pose respiratory risks to nearby communities [12,16,17]. Experimental research has demonstrated that wind erosion of dry tailings surfaces leads to increased emissions of particulate matter [12,18]. These examples underscore the critical need for effective erosion control measures and the careful design of tailings disposal sites to protect the environment and public health.

Scientific findings provide various methods for the surface stabilization of flotation tailings disposal sites to mitigate wind erosion. A range of biological, chemical, and physical approaches to stabilize the surfaces of flotation tailings disposal sites, thereby reducing the impact of wind erosion and improving environmental safety, has been applied. In addition to the application of *Bacillus sphaericus* for soil surface stabilization [19], or indigenous bacteria isolated from tailings soils with the potential to form biocement [20], volcanic ash-based geopolymers have shown notable performance in stabilizing silty and sandy soils [21]. Natural plant covers have been identified as an optimal and sustainable method of preventing and controlling wind erosion on flotation tailings disposal sites [15,22,23]. Among many plant species, grasses have been studied for their effectiveness in tailings stabilization, wind erosion reduction, and biomass accumulation during their growth [22,24,25].

The key characteristics of grass species that make them suitable for the sustainable reclamation of degraded land include their adaptation to extreme habitat conditions and their ability to increase biomass production. However, the challenges and limitations associated with using grass species for land reclamation include the possible negative impact of certain grass species on species richness and diversity, as well as the potential for establishment limitations hindering the recolonization of target species. Therefore, careful consideration of environmental suitability and proper management practices is essential for successful reclamation using perennial grasses. The increase in the abundance of certain grass species, including invasive species, can have a negative impact on species richness and diversity, as well as soil structure and moisture content, highlighting the need for the careful selection of grass species for reclamation [26].

This research aimed to determine the feasibility of using grass species for the sustainable restoration of natural vegetation cover on flotation tailings from an unrehabilitated post-mining heap. To our current knowledge, this is the first research report testing plant survival on pure flotation tailing as a growing medium.

2. Materials and Methods

The study used flotation tailings from a copper ore mine near Polkowice, Poland (51.520 N, 16.200 E). The tailings were taken from the tailings beach boundaries (sample 0–20 cm deep), which were dry and had a finely fragmented structure. River sand was used as the control. The analysis of both substrates was performed at the Regional Chemical-Agricultural Station in Warsaw, from the moment of sample collection until the presentation of the results (Table 1).

Table 1. The physico-chemical characteristics of the substrate used in the experiment and the procedure applied.

Parameter	Substrate:		Accredited Procedure:
	Flotation Tailing	Sand (Control)	
Soil composition:			laser diffraction method
sand (0.05–2.0 mm)	94.96	100.0	
silt (0.002–0.05 mm)	5.04	0.0	
clay (<0.002 mm)	0	0.0	
organic matter content (%)	<0.30 *	<0.30 *	titration method
pH (KCl)	8.4±	8.1	potentiometric method
Macro- and micronutrients (in mg kg^{-1})			
N-NO_3	<5 *	<5 *	Continuous Flow Analysis (CFA) method with spectrophotometric detection
N-NH_4	5	8	
Cl	<10 *	11	potentiometric method
P	23	5	Flame Atomic Absorption Spectrometry (FAAS) method
K	20	10	
Ca	5350	440	
Mg	60	25	
Cu	24.0	1.1	
Fe	13.4	15.2	
Mn	9.2	9.9	
Zn	1.8	3.9	

Explanation: *—result below the limit of quantification (LOQ).

Eight grass species were used as plant material for the study. Three of them were cultivars commonly used for forage or lawn purposes with a significant share of the grass seed market (tall fescue (*Festuca arundinacea* Schreb., FA) cv. BARFELIX, red fescue (*Festuca rubra* L., FR) cv. ARETA, and perennial ryegrass (*Lolium perenne* L., LP) cv. BOKSER. The next two were forage forms from the genus brome: smooth brome (*Bromus inermis* Leyss. BI) cv. LINCOLN, and rescue grass (*Bromus catharticus* Vahl., BC) cv. BROMA. The next one was a cultivar of tall wheatgrass (*Elytrigia elongata* Vahl., EE) cv. BAMAR, a species confirmed to be suitable for growing in saline and alkaline areas. The remaining two were grass species commonly found in degraded, ruderal areas: annual species-drooping brome or cheatgrass (*Bromus tectorum* L., BT) and velvet grass (*Holcus lanatus* L., HL). The seed of the above-mentioned cultivars were kindly provided by breeders and the seeds of the latter two species were collected from ruderal areas by authors.

2.1. Seed Germination Test

In March 2022, seeds of the grass varieties mentioned above were sown in the above-described substrates. Plastic trays (12 × 23 × 3 cm) were filled with sand or flotation tailings with a layer of about 3 cm thick. For each cultivar and each type of substrate, 2 trays were used and 50 seeds of each cultivar were placed in the substrate. Then, after watering with distilled water, trays were covered with foil to avoid the rapid drying of the substrate. After this, they were kept in air-conditioned glasshouse with the temperature set at 25–30 °C (daytime) and 10–15 °C (nighttime), with a 12 h day/night cycle. Substrates were sprayed with water at 2-day intervals. The experiment was repeated for 2 runs.

After 5 and 14 days, seedlings classified as normal were counted. Further, seedlings in trays were irrigated with 0.02% KNO_2 for a further 5 days to ensure that all potentially dormant seeds germinated.

At the same time, to assess the germination capacity, seeds of the same varieties were sown on a Jacobsen-type germination device (3 × 100 seed per cultivar), using the lighting and temperature parameters as above, in line with the ISTA recommendations [27]. Three hundred seeds (3 × 100) from each selected cultivar were placed on filter paper on the germination device, with light and temperature regimes as provided above. After 5 and 14 days, seedling classification and counting were performed.

2.2. Seedling Growth

Seeds of the above-mentioned grass species were placed in test substrates in metal pots (50 seeds per pot, 3 pots per species per each substrate, 5000 cm^3, 20 cm in diameter, with a drainage hole in the bottom). Pots were further also kept in an air-conditioned glasshouse with temperature set at 25–30 °C (daytime) and 10–15 °C (nighttime), with a 12 h day/night cycle. Pots were sprayed with water at 2-day intervals. This experiment was also repeated for 2 runs. The height of the seedlings was assessed after 14 days of seedling growth in test substrates, and after 20 days of growth the length of the root system after removing the plants from the substrate was measured.

2.3. Statistical Analysis

All statistical calculations were made with STATISTCA ver. 12.0 PL. Mean values from two runs were used for further calculations, as there were no significant differences between runs. The significance of differences was determined with 95% probability. Least significant differences (LSD) were calculated according to the Fisher test and values were shown only if statistically significant with accepted probability. To calculate the significance of the difference between germination in sand and flotation tailings, T-test was used. Homogeneity groups were determined using the Duncan test, with the probability of difference between means being greater than 95%.

3. Results

3.1. Germination Test

Seeds of all the studied species were germinated above 80% under ISTA-compliant conditions (Table 2). On the other hand, germination in sand and flotation tailings indicated considerable variability among the studied forms. Notably, species belonging to the *Bromus* genus exhibited superior germination rates in sand compared to the germination tray used in the ISTA test and in the flotation tailings. This observation was particularly evident in the case of *Bromus tectorum* (BT), which demonstrated equally impressive germination rates in flotation tailings (93.5%). Among the studied species, only BT, *Bromus catharticus* (BC), and *Elyrigia elongata* (EE) achieved germination rates of at least 80% in flotation tailings. The remaining species exhibited germination rates below 60% in this medium. *Festuca rubra* (FR) and *Lolium perenne* (LP), two species commonly utilized for sowing lawns, roadsides, and wasteland due to their rapid emergence and sod formation, were identified as the poorest germinators in flotation tailings (31.8 and 26.3%, respectively).

Table 2. Seed germination rates (%) of eight grass species under three different conditions.

Genus Species	Germination: acc. to the ISTA Rules	In Sand	In Flotation Tailing
	% [Mean ± std.dev.]		
Bromus inermis LINCOLN	83.3 ± 1.51 B b	93.3 ± 4.08 A a	57.2 ± 3.99 C cd
Bromus tectorum, ecotype	85.5 ± 1.87 B ab	93.0 ± 3.58 A a	93.5 ± 3.08 A a
Bromus catharticus BROMA	80.6 ± 7.37 AB b	91.7 ± 6.38 A ab	81.2 ± 10.17 AB ab
Elytrigia elongata BAMAR	94.0 ± 1.55 A a	91.0 ± 3.58 A ab	79.7 ± 9.33 B b
Festuca arundinacea BARFELIX	87.5 ± 4.10 A ab	54.0 ± 4.81 B e	60.7 ± 8.61 B c
Festuca rubra ARETA	83.2 ± 1.47 A b	69.0 ± 9.96 B c	31.8 ± 1.94 C e
Holcus lanatus, ecotype	84.2 ± 3.49 A ab	61.0 ± 3.03 B d	52.2 ± 2.13 C d
Lolium perenne BOKSER	90.0 ± 3.90 A a	34.1 ± 3.39 B f	26.3 ± 0.75 C f

Explanation: Different lowercase letters (a, ab, b, etc.) indicate statistically significant differences between grass species, while different uppercase letters (A, AB, etc.) show differences between germination medium (Duncan's test, $p > 95\%$).

3.2. Seedling Growth and Development

In general, grass seedlings grown in flotation tailing were ca. 32% shorter than seedlings grown in pure sand (Table 3, Figure 1).

Table 3. Results of measurements of seedling characteristics for grass species grown in two different growing substrates.

Genus Species	Growing Substrate	Seedling Height [cm] Mean ± std.dev.	Seedling Height [cm] T-Test	Length of the Root System [cm] Mean ± std.dev.	Length of the Root System [cm] T-Test
Bromus inermis LINCOLN	Sand flot. tailing	6.2 ± 1.47 5.8 ± 1.17	0.43 ns	8.4 ± 0.80 13.0 ± 1.41	−6.91 ***
Bromus tectorum ecotype	Sand flot. tailing	7.0 ± 1.90 6.3 ± 1.08	0.84 ns	11.3 ± 1.21 24.3 ± 1.63	−15.66 ***
Bromus catharticus BROMA	Sand flot. tailing	11.2 ± 1.17 5.0 ± 1.55	7.78 ***	9.5 ± 1.05 12.7 ± 1.97	−3.48 ***
Elytrigia elongata BAMAR	Sand flot. tailing	12.8 ± 1.94 9.7 ± 1.21	3.39 **	22.0 ± 1.41 19.0 ± 1.09	4.11 ***
Festuca arundinacea BARFELIX	Sand flot. tailing	5.2 ± 1.47 3.2 ± 1.40	2.31 **	4.3 ± 1.75 2.5 ± 1.05	2.20 ns
Festuca rubra ARETA	Sand flot. tailing	4.7 ± 1.03 2.9 ± 1.02	2.95 **	1.5 ± 1.05 0.8 ± 0.26	1.51 ns
Holcus lanatus ecotype	Sand flot. tailing	2.6 ± 1.43 1.5 ± 1.00	1.52 ns	2.2 ± 1.17 0.8 ± 0.26	2.73 **
Lolium perenne BOKSER	Sand flot. tailing	4.0 ± 0.89 1.4 ± 0.66	5.68 ***	2.3 ± 0.82 0.8 ± 0.26	4.29 ***
Mean	Sand flot. tailing	6.3 ± 1.42 4.3 ± 1.09	3.32 ***	7.6 ± 1.11 8.8 ± 0.97	0.82 ns
Difference between species in substrates used:	Sand flot. tailing	*** ***		*** ***	

Explanation: designation of statistical significance of the difference between means: ns—no significant difference; ***—significance with $p > 0.001$; **—significance with $p > 0.05$.

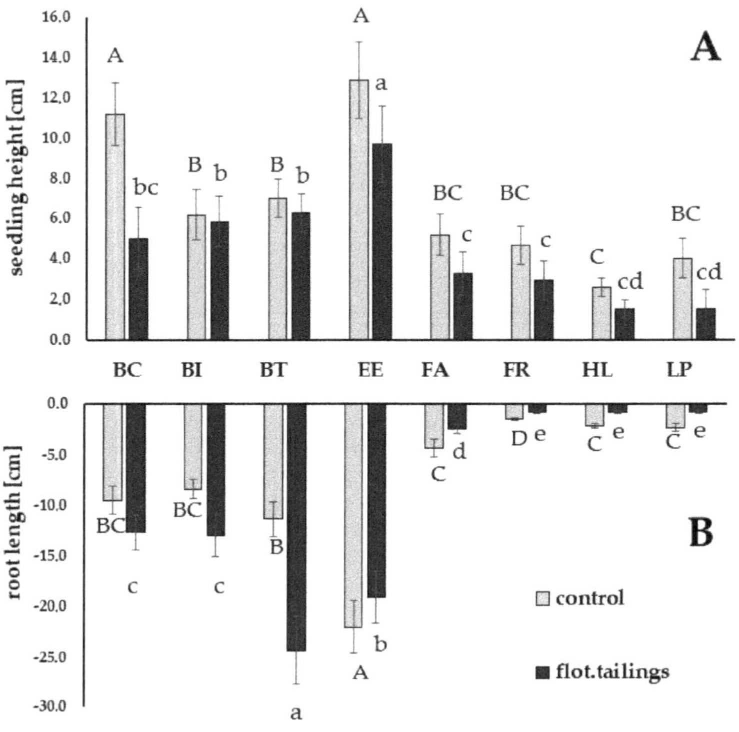

Figure 1. Differences in seedling height (**A**) and seedling root length (**B**) of grass species grown in flotation tailings and control media. Upper-case letters (A, B, BC, C, D) refer to homogeneity groups (p > 95%) acc to Duncan test for control substrate; lower-case letters (a, b, bc, c, d, cd, e) refer to flotation tailings as a substrate. Bars indicate standard deviation from the mean value. Species code: BC—*Bromus carinathus;* BI—*Bromus inermis;* BT—*Bromus tectorum;* EE—*Elytrigia elongata;* FA—*Festuca arundinacea;* FR—*Festuca rubra;* HL—*Holcus lanatus;* LP—*Lolium perenne.*

In the case of *Bromus inermis* (BI), BT and *Holcus lanatus* (HL) no statistical difference was calculated between seedling height measured in both growing media. The highest reduction in seedling height grown in flotation tailings (ca. 65%) was noted for LP.

Seedlings root development in flotation tailings was promoted for grass species from genus *Bromus,* while it was significantly reduced for EE, HL and LP. The length of seedling roots grown in flotation tailings as compared to those grown in sand was 215% for BT, 154% for BI, and 133% for BC.

Despite the statistically significant effect of the seedling root length on the reduction in EE when grown in flotation tailings as compared to control, the previously mentioned species performed much better than FA, FR, and LP. Within the time frame provided in the experiment, during the growth in flotation tailings, EE produced the longest seedlings with the second (after BT) longest roots.

The seed germination of the investigated cultivars in test substrates was, as anticipated, strongly conditioned by the initial germination, which was conducted following the ISTA rules (Table 4).

Notably, the highest, positive, and statistically significant Pearson correlation coefficient was observed between germination in sand and germination in flotation tailings ($r = 0.89$, $p > 0.001$). Nevertheless, no significant correlation was observed between germination test results according to ISTA standards and seedling growth parameters in the tested substrates. Seed germination in sand was a good predictor of seedling growth in flotation tailings; however, this correlation did not extend to sand substrate.

Table 4. Results of the Pearson correlation analysis between seed germination under different conditions and seedling growth in two different growing media.

Germination Test:	Germination Test in:		Growing Substrate:			
	Sand	Flotation Tailings	Sand		Flotation Tailing	
			Seedling Height	Length of Roots	Seedling Height	Length of Roots
ISTA rules	0.89 ***	0.78 **	0.43 ^{ns}	0.49 ^{ns}	0.59 ^{ns}	0.52 ^{ns}
sand	x	0.88 **	0.59 ^{ns}	0.60 ^{ns}	0.71 **	0.74 **
flot.tailings	x	x	0.67 ^{ns}	0.72 **	0.74 **	0.87 **

Explanation: designation of statistical significance of the Pearson correlation coefficients: ns—no significant; ***—significance with $p > 0.001$; **—significance with $p > 0.05$.

4. Discussion

Seed germination in laboratory conditions tends to be more controlled and predictable, while, in contrast, germination in sand or soil is subject to a range of environmental and biotic factors that can influence the outcomes, making it more representative of natural conditions [28,29]. Our research has shown that seed germination in laboratory conditions, such as Petri dishes, can sometimes be lower than that in natural substrates like sand or soil. In this experiment, grasses from the genus Bromus were used. Generally, germination in sand is a reliable predictor of germination in tailings, but germination according to ISTA recommendations is not helpful for this specific application.

For example, studies on the seeds of desert plants such as *Prosopis juliflora* and *Acacia nilotica* demonstrated that germination rates are often higher in sandy soils compared to controlled laboratory conditions [30,31]. This is primarily because natural substrates provide more consistent moisture retention, better aeration, and appropriate microflora that enhance germination, conditions that are challenging to replicate accurately in a laboratory setting.

The species that is potentially most useful for restoring vegetation on a post-flotation sediment landfill should be characterized by the ability to germinate rapidly in such a substrate and to effectively continue the initial growth, i.e., produce seedlings with a developed root system. In light of the obtained results, *Bromus inermis*, *B. catharticus*, and *Elytrigia elongata* can be considered promising species.

Bromus inermis (smooth bromegrass), has been described as the most important grass for soil erosion purposes in the north central states of USA [32]. As a cool-season, rhizomatous grass of Eurasian origin, smooth bromegrass escaped from intentional plantings in North America and spread widely in natural areas [33]. It is recommended for the reclamation of open-pit mines and fly ash landfills [33,34]. Its extensive root system promotes adaptation to various soil types and environments, including dryland, weak and wet soils, waterlogged sites, embankments, and escarpments [35].

B. catharticus is a valuable grass, supporting livestock production through high-quality forage and seed, while also contributing to soil and water conservation and improved pasture health due to its robust growth [36,37]. Due to its strong regeneration and ability to adapt to diverse environments, *B. catharticus* holds potential for soil and water conservation and cultivated pasture improvement [38]. However, information concerning the suitability of this species for reclamation is lacking.

Last, of the mentioned species, tall wheatgrass (*Elytrigia elongata*) exhibits exceptional tolerance to soil salinity and alkalinity, withstanding soluble salt concentrations of up to 1%. In its natural habitat, it is often found in saline or alkaline soils [39]. This plant thrives in moderately to severely saline areas and tolerates winter waterlogging and summer drought in both acidic and alkaline soils [40–42]. Our experiment also confirmed that tall wheatgrass produces extensive root systems that help to stabilize soil and prevent erosion [43].

The next species that germinated and grew very well in flotation sediments was *Bromus tectorum* (BT). However, its potential usefulness in revegetation processes requires further in-depth analysis. The superior germination and seedling development of BT could be the combined effect of the nature of this species (annual species) and other factors. Adaptive germination responses in BT populations in contrasting habitats may have both genetic and environmental components, thus explaining why this species can become established in such a variety of habitats [44]. It has been suggested that, on the basis of the germination responses of seed from contrasting environments, BT, together with *Bromus arvensis*, seem to require a disturbed habitat to grow and establish a population, in contrast to an agricultural landscape, where they are quite infrequent [45]. Annuals, such as BT, often exhibit strategies such as a rapid growth, early maturation, and high seed production, which allows them to quickly utilize the available resources in unstable environments [46,47]. In such environments, natural selection favors organisms that reproduce and spread rapidly.

On the other hand, BT negatively impacts grass and shrublands in arid and semiarid regions, outcompeting native species and leading to shifts in ecosystem function and plant community composition [47–49]. Despite *Bromus tectorum* obtaining the best results regarding germination and initial seedling growth in flotation tailings, considering the aggressive and invasive nature of this species, no practical suggestions related to its use in reclamation issues can be made [50,51]. The result of the flotation tailings reclamation should not be the introduction of potentially aggressive weeds that can easily spread to areas adjacent to the reclaimed sites, thereby hindering ecological restoration efforts [52,53].

Grass varieties used in our experiment, such as *Festuca arundinacea* cv BARFELIX, *Festuca rubra* cv ARETA, and *Lolium perenne* cv BOKSER, were found to exhibit pronounced sensitivity to flotation tailings as a substrate for initial growth and development. The experimental evidence indicates that the conditions associated with flotation tailings, which are characterized by chemical and physical properties that deviate from those of conventional soils, adversely affect both the germination processes and the subsequent growth dynamics of plants [7,54,55]. The germination rates under these suboptimal conditions were significantly lower than expected, and the subsequent developmental stages failed to show the vigor typically observed in more conducive environments. Similar conclusions concerning the unsuitability of *Lolium perenne* to germinate and grow on flotation tailings were suggested by Jelea and Baciu [15]. Such an effect could be the result of the inhibitory factors of a physical and chemical nature (pH, Eh, heavy metals), as well as the granulometry of tailings (germination layer density, air presence, capillary force, etc.).

5. Conclusions

A comprehensive understanding of the biological growth and development of vegetation targeted for reclamation, in areas contaminated by, e.g., flotation sediments, is crucial for the strategic planning of sustainable reclamation initiatives. This understanding allows for the leveraging of natural processes, which culminates in the creation of a durable vegetative layer, preventing both water and wind erosion. The selection of suitable plant varieties, including grasses, stands as a cornerstone in determining the efficacy of the reclamation endeavor. The results of the studies documented herein identify a mere three cultivars of three perennial grass species demonstrating potential efficacy for the restoration of natural vegetation cover on a substrate of flotation tailings originating from an unrehabilitated heap: *Bromus inermis* cv. LINCOLN, *Bromus catharticus* cv. BROMA, and *Elytrigia elongata* cv. BAMAR.

This sensitivity raises concerns about the viability and effectiveness of these species in practical field applications, where soil quality is compromised by industrial residues. Consequently, the revegetation of post-industrial landscapes with flotation tailings as a primary medium for the restoration may require additional amendments or alternative species selections to achieve successful vegetative cover and long-term ecosystem stabilization. By evaluating a range of grass species directly on the challenging substrate of copper flotation tailings, this study provides valuable methodological insights for selecting appropriate vegetation that can not only tolerate harsh conditions but will also contribute to the long-term stability and ecological integrity of the restored landscape.

The problem of flotation tailings' stabilization and reclamation is a microcosm of the larger environmental sustainability challenge. It highlights the interconnectedness of soil degradation, biodiversity loss, environmental contamination, and unsustainable industrial practices. The research being conducted represents an effort to mitigate these negative impacts and restore degraded ecosystems, emphasizing the importance of finding sustainable solutions to industrial waste management.

Future research could further explore the physiological mechanisms underlying the tolerance of successful species and investigate their impact on soil microbial communities and nutrient cycling, thereby deepening our understanding of sustainable soil reclamation in such specific degraded environments [56].

Author Contributions: Conceptualization: G.Ż. and D.M., methodology: G.Ż. and D.M.; formal analysis: D.M.; investigation: D.M.; resources: G.Ż.; data curation: G.Ż. and D.M.; writing—original draft preparation: G.Ż.; writing—review and editing: G.Ż. All authors have read and agreed to the published version of the manuscript.

Funding: This research received no external founding.

Institutional Review Board Statement: Not applicable.

Informed Consent Statement: Not applicable.

Data Availability Statement: The original contributions presented in this study are included in the article. Further inquiries can be directed to the corresponding author.

Conflicts of Interest: The authors declare no conflicts of interest.

References

1. Leopold, A. *A Sand County Almanac and Sketches Here and There*; Penguin Random House: London, UK, 2020; p. 173.
2. Oldeman, L.R. Global Extent of Soil Degradation. In *ISRIC Bi-Annual Report 1991–1992*; International Soil Reference and Information Centre: Wageningen, The Netherlands, 1992; pp. 19–36.
3. Baumhardt, R.L.; Stewart, B.A.; Sainju, U.M. North American Soil Degradation: Processes, Practices, and Mitigating Strategies. *Sustainability* **2015**, *7*, 2936–2960. [CrossRef]
4. Biegańska, J. Remediation and revitalization of industrial areas—A method of promotion. *IOP Conf. Ser. Mater. Sci. Eng.* **2019**, *545*, 012002. [CrossRef]
5. Pasieczna, A. Soil contamination induced by historical zinc-lead ore mining and iron and zinc smelting in the central part of the upper silesian industrial region (Southern Poland). *Biul. Państwowego Inst. Geol.* **2018**, *473*, 49–66. [CrossRef]
6. Pasieczna, A.; Konon, A.; Nadłonek, W. Sources of anthropogenic contamination of soil in the upper silesian agglomeration (Southern Poland). *Geol. Quaterly* **2020**, *64*, 988–1003. [CrossRef]
7. Gondek, K.; Mierzwa-Hersztek, M.; Kopeć, M.; Bajda, T. Phytostabilisation on post-flotation sediment waste: Mobility of heavy metals and stimulation of biochemical processes by mineral-organic mixtures. *J. Soils Sediments* **2020**, *20*, 3502–3513. [CrossRef]
8. Michalski, K. Analysis of the Features of a Degraded Land in Terms of Revitalization—Case Study on the Example of a Post-Smelter Area. *Multidiscip. Asp. Prod. Eng. MAPE* **2020**, *3*, 559–569. [CrossRef]

9. Ciesiółka, P. Urban regeneration as a new trend in the development policy in Poland. *Quaestionnes Geogr.* **2018**, *37*, 109–123. [CrossRef]
10. Hilson, G. An overview of land use conflicts in mining communities. *Land Use Policy* **2002**, *19*, 65–73. [CrossRef]
11. Pietrzyk-Sokólska, E.; Uberman, R.; Kulczycka, J. Wpływ górnictwa na środowisko w Polsce—Mity i rzeczywistość. *Gospod. Surowcami Miner. Miner. Resour. Manag.* **2015**, *31*, 45–64. [CrossRef]
12. Cacciuttolo, C.; Cano, D.; Custodio, M. Socio-Environmental Risks Linked with Mine Tailings Chemical Composition: Promoting Responsible and Safe Mine Tailings Management Considering Copper and Gold Mining Experiences from Chile and Peru. *Toxics* **2023**, *11*, 462. [CrossRef]
13. Łuszczkiewicz, A. Koncepcje wykorzystania odpadów flotacyjnych z przeróbki rud miedzi w regionie legnicko-głogowskim. *Inżynieria Miner.* **2000**, *1*, 25–35.
14. Haneef Mian, M.; Yanful, E. Tailings erosion and resuspension in two mine tailings ponds due to wind waves. *Adv. Environ. Res.* **2003**, *7*, 745–765. [CrossRef]
15. Jelea, O.-C.; Baciu, C. Effects of heavy metals, contained in flotation tailings, on plants used in revegetation. *Carpathian J. Earth Environ. Sci.* **2023**, *18*, 89–103. [CrossRef]
16. Slingerland, N.; Sommerville, A.; O'Leary, D.; Beier, N.A. Identification and quantification of erosion on a sand tailings dam. *Geosystem Eng.* **2018**, *23*, 131–145. [CrossRef]
17. van Wyk, S.J.; Hatting, J.; Haagner, A.S.H. Wind erosion design considerations for closure of tailings storage facilities in South Africa: A case study. In *Mine Closure 2019: Proceedings of the 13th International Conference on Mine Closure, Australian Centre for Geomechanics, Perth*; ACG: Crawley, Australia, 2019; pp. 1185–1200. [CrossRef]
18. Dentoni, V.; Grosso, B.; Pinna, F. Experimental evaluation of PM emission from Red Mud Basins exposed to wind erosion. *Minerals* **2021**, *11*, 405. [CrossRef]
19. Namdar-Khojasteh, D.; Bazgir, M.; Hashemi Babaheidari, S.A.; Akwasi BAsmudu-Sakyi, A.B. Application of biocementation technique using *Bacillus sphaericus* for stabilization of soil surface and dust storm control. *J. Arid Land* **2002**, *14*, 537–549. [CrossRef]
20. Maureira, A.; Zapata, M.; Olave, J.; Jeison, D.; Wong, L.-S.; Panico, A.; Hernández, P.; Cisternas, L.A.; Rivas, M. MICP mediated by indigenous bacteria isolated from tailings for biocementation for reduction of wind erosion. *Front. Bioeng. Biotechnol.* **2024**, *12*, 1393334. [CrossRef]
21. Mohebbi, H.R.; Javadi, A.A.; Azizkandi, A.S. The Effects of Soil Porosity and Mix Design of Volcanic Ash-Based Geopolymer on the Surface Strength of Highly Wind Erodible Soils. *Minerals* **2022**, *12*, 984. [CrossRef]
22. Andrejić, G.; Kovaćević, M.; Dzeletović, Z.; Aleksić, U.; Grdović, I.; Rakić, T. Potentially toxic element accumulation in two *Equisetum* species spontaneously grown in the flotation tailings. *J. Serbian Chem. Soc.* **2023**, *88*, 1055–1064. [CrossRef]
23. Zhang, J.; Jia, Z.; Li, Q.; He, L.; Zhao, X.; Wang, L.; Han, D. Determine the Optimal Vegetation Type for Soil Wind Erosion Prevention and Control in the Alpine Sandy Land of the Gonghe Basin on the Qinghai Tibet Plateau. *Forests* **2023**, *14*, 2342. [CrossRef]
24. Antonelli, P.M.; Coghill, M.G.; Gardner, W.C.; Fraser, L.H. Semiarid bunchgrasses accumulate molybdenum on alkaline copper mine tailings: Assessing phytostabilization in the greenhouse. *SN Appl. Sci.* **2021**, *3*, 747. [CrossRef]
25. El Berkaoui, M.; El Adnani, M.; Hakkou, R.; Ouhammou, A.; Bendaou, N.; Smouni, A. Assessment of the Transfer of Trace Metals to Spontaneous Plants on Abandoned Pyrrhotite Mine: Potential Application for Phytostabilization of Phosphate Wastes. *Plants* **2022**, *11*, 179. [CrossRef] [PubMed]
26. Puffer, M.; Limb, R.F.; Daigh, A.L.M.; Sedivec, K.K. Mechanical and biotic reclamation strategies for a post-mine temperate grassland. *Land Degrad. Dev.* **2023**, *34*, 3114–3129. [CrossRef]
27. ISTA. *International Rules for Seed Testing*; The International Seed Testing Association: Wallisellen, Switzerland, 2023. [CrossRef]
28. Baskin, C.C.; Baskin, J.M. *Seeds: Ecology, Biogeography, and, Evolution of Dormancy and Germination*, 2nd ed.; Academic Press, Elsevier Inc.: Amsterdam, The Netherlands, 2014; p. 1586. [CrossRef]
29. Bewley, J.D.; Bradford, K.J.; Hilhorst, H.W.; Nonogaki, H. *Seeds: Physiology of Development, Germination and Dormancy*; Springer Science & Business Media: Berlin/Heidelberg, Germany, 2013; p. 392. [CrossRef]
30. Nascimento, C.E.d.S.; da Silva, C.A.D.; Leal, I.R.; Tavares, W.d.S.; Serrão, J.E.; Zanuncio, J.C.; Tabarelli, M. Seed germination and early seedling survival of the invasive species *Prosopis juliflora* (Fabaceae) depend on habitat and seed dispersal mode in the Caatinga dry forest. *PeerJ* **2020**, *8*, e9607. [CrossRef]
31. Gilani, M.M.; Ahmad, I.; Farooq, T.H.; Wu, P.; Yousaf, M.S.; Khan, M.W.; Yousuf Bin, T.; Ma, X. Effects of pre-sowing treatments on seed germination and morphological growth of *Acacia nilotica* and *Faidherbia albida*. *Sci. For. Piracicba* **2019**, *47*, 374–382. [CrossRef]
32. Vogel, K.P.; Moore, K.J.; Moser, L.E. Bromegrass. In *Cool-Season Forage Grasses*; Agronomy Monograph 34; Moser, L.E., Buxton, D.R., Casler, M.D., Eds.; American Society of Agronomy, Crop Science Society of America, Soil Science Society of America: Madison, MI, USA, 1996; pp. 535–567.
33. Salesman, J.B.; Thomsen, M. Smooth Brome (*Bromus inermis*) in Tallgrass Prairies: A Review of Control Methods and Future Research Directions. *Ecol. Restor.* **2011**, *29*, 374–381. [CrossRef]

34. Klimont, K.; Bulińska-Radomska, Z. Badanie Rozwoju Wybranych Gatunków Traw Do Umacniania Składowisk Popiołów Paleniskowych z Elektrociepłowni. *Probl. Inżynierii Rol.* **2009**, *17*, 135–145.
35. Raawe, H. About Grass and Legume Species Suitability for Recultivation of Semi-Coke Dumps. *Agronomy* **2004**, *219*, 154–156.
36. Hume, D.E. Primary growth and quality characteristics of *Bromus willdenowii* and *Lolium multiflorum*. *Grass Forage Sci.* **1990**, *46*, 313–326. [CrossRef]
37. Yi, L.; Dong, Z.; Lei, Y.; Zhao, J.; Xiong, Y.; Yang, J.; Xiong, Y.; Gou, W.; Ma, X. Genetic Diversity and Molecular Characterization of Worldwide Prairie Grass (*Bromus catharticus* Vahl) Accessions Using SRAP Markers. *Agronomy* **2021**, *11*, 2054. [CrossRef]
38. Belesky, D.P.; Ruckle, J.M.; Abaye, A.O. Seasonal distribution of herbage mass and nutritive value of Prairiegrass (*Bromus catharticus* Vahl). *Grass Forage Sci.* **2007**, *62*, 301–311. [CrossRef]
39. Asay, K.H.; Jensen, K.B. Wheatgrasses. In *Cool-Season Forage Grasses*; Agronomy Monograph 34; Moser, L.E., Buxton, D.R., Casler, M.D., Eds.; American Society of Agronomy, Crop Science Society of America, Soil Science Society of America: Madison, MI, USA, 1996; pp. 691–724.
40. Taboada, M.A.; Rubio, G.; Lavado, R. The deterioration of tall wheatgrass pastures on saline sodic soils. *J. Rangel. Manag.* **1998**, *51*, 241–246. [CrossRef]
41. Nichols, C. *Establishing and Managing Tall Wheat Grass in Saline Soils for Productivity*; Agricultural Notes; VGLS: Werribee, Australia, 1998; pp. 1–2.
42. Csete, S.; Stranczinger, S.; Szalontai, B.; Farkas, A. Tall wheatgrass cultivar Szarvasi-1 (*Elymus elongatus* subsp. *ponticus* cv. Szarvasi-1) as a potential energy crop for semi-arid lands of Eastern Europe. In *Sustainable Growth and Applications in Renewable Energy Sources*; Nayeripour, M., Kheshti, M., Eds.; InTech: Rijeka, Croatia, 2011; pp. 269–294.
43. Vergiev, S. Comparative Study of the capacity of three plant species from the Poaceae family for erosion and flooding control of coastal areas. Sustainable Development and Innovations in Marine Technologies. In Proceedings of the 18th International Congress of the International Maritime Association of the Mediterranean, IMAM 2019, Varna, Bulgaria, 9–11 September 2019; pp. 597–602.
44. Beckstead, J.; Meyer, S.E.; Allen, P.S. *Bromus tectorum* seed germination: Between-population and between-year variation. *Can. J. Bot.* **1996**, *75*, 875–882. [CrossRef]
45. Andersson, L.; Milberg, P.; Schütz, W.; Steinmetz, O. Germination characteristics and emergence time of annual Bromus species of differing weediness in Sweden. *Weed Res.* **2002**, *42*, 135–147. [CrossRef]
46. Franzese, J.; di Virgilio, A.; Pirk, G.; Lescano, M.N.; Speziale, K.L. Low biotic resistance to cheatgrass invasion in Patagonia: Evidence from competition experiments. *Biol. Invasions* **2022**, *24*, 235–246. [CrossRef]
47. Howel, A.; Winkler, D.E.; Phillips, M.L.; McNellis, B.; Reed, S.C. Experimental Warming Changes Phenology and Shortens Growing Season of the Dominant Invasive Plant *Bromus tectorum* (Cheatgrass). *Front. Plant Sci.* **2020**, *11*, 570001. [CrossRef]
48. Mack, R.N. Invasion of *Bromus tectorum* L. into Western North America: An ecological chronicle. *Agro-Ecosystems* **1981**, *7*, 145–165. [CrossRef]
49. Emam, T.M.; Espeland, E.K.; Rinella, M.J. Soil sterilization alters interactions between the native grass *Bouteloua gracilis* and invasive *Bromus tectorum*. *J. Arid. Environ.* **2014**, *111*, 91–97. [CrossRef]
50. Seastedt, T.R.; Pysek, P. Mechanisms of plant invasions of North American and European grasslands. *Annu. Rev. Ecol. Evol. Syst.* **2011**, *42*, 133–153. [CrossRef]
51. Pyšek, P.; Manceur, A.M.; Alba, C.; McGregor, K.F.; Pergl, J.; Štajerová, K.; Chytrý, M.; Danihelka, J.; Kartesz, J.; Klimešová, J.; et al. Naturalization of central European plants in North America: Species traits, habitats, propagule pressure, residence time. *Ecology* **2015**, *96*, 762–774. [CrossRef]
52. Sheley, R.L.; Krueger-Mangold, J. Principles for restoring invasive plant-infested rangeland. *Weed Sci.* **2003**, *51*, 260–265. Available online: http://www.jstor.org/stable/4046729 (accessed on 15 January 2025). [CrossRef]
53. Hulme, P.E.; Bacher, S.; Kenis, M.; Klotz, S.; Kühn, I.; Minchin, L.; Nentwig, W.; Olenin, S.; Panov, J.; Pergl, J.; et al. Grasping at the routes of biological invasions: A framework for integrated biosecurity. *J. Appl. Ecol.* **2008**, *45*, 403–414. [CrossRef]
54. Wong, M.H. Ecological restoration of mine degraded soils, with emphasis on metal contaminated soils. *Chemosphere* **2003**, *50*, 775–780. [CrossRef] [PubMed]
55. Krawczyńska, M.; Kołwzan, B.; Gediga, K.; Dziubek, A.M.; Grabas, K.; Karpenko, E. Evaluation of the possibility of phytostabilization of post-flotation tailings pond. *Environ. Prot. Eng.* **2015**, *41*, 157–167. [CrossRef]
56. van der Putten, W.H.; Bardgett, R.D.; Bever, J.D.; Bezemer, T.M.; Casper, B.B.; Fukami, T.; Wardle, D.A. Plant-soil feedback cycles as drivers of plant community assembly. *Trends Ecol. Evol.* **2016**, *31*, 589–601.

MDPI AG
Grosspeteranlage 5
4052 Basel
Switzerland
Tel.: +41 61 683 77 34

Sustainability Editorial Office
E-mail: sustainability@mdpi.com
www.mdpi.com/journal/sustainability

www.ingramcontent.com/pod-product-compliance
Lightning Source LLC
Chambersburg PA
CBHW061626210926
51850CB00009B/2452
* 9 7 8 3 7 2 5 8 5 5 7 3 5 *